Satoyama Initiative Thematic Review

Series Editor

Tsunao Watanabe, United Nations University Institute for the Advanced Study of Sustainability (UNU-IAS), Tokyo, Japan

This Open Access book series aims to make timely and targeted contributions for decision-makers and on-the-ground practitioners by producing knowledge concerning "socio-ecological production landscapes and seascapes" (SEPLS) – areas where production activities help maintain biodiversity and ecosystem services in various forms while sustainably supporting the livelihoods and well-being of local communities. Each volume will be designed as a compilation of case studies providing useful knowledge and lessons focusing on a specific theme that is important for SEPLS, accompanied by a synthesis chapter that extracts lessons learned through the case studies to present them for policy-relevant academic discussions. The series is also intended to contribute to efforts being made by researchers and scientists to strengthen the evidence base on socio-ecological dynamics and resilience, including those under the Intergovernmental Science-Policy Platform on Biodiversity and Ecosystem Services (IPBES) and the Convention on Biological Diversity (CBD).

The promotion and conservation of SEPLS have been the focus of the Satoyama Initiative, a global effort to realise societies in harmony with nature. In 2010, the International Partnership for the Satoyama Initiative (IPSI) was established to implement the concept of the Satoyama Initiative and promote various activities by enhancing awareness and creating synergies among those working with SEPLS. As a unique platform joined by governmental, intergovernmental, nongovernmental, private-sector, academic, and indigenous peoples' organisations, IPSI has been collecting and sharing the information, lessons, and experiences on SEPLS to accumulate a diverse range of knowledge. Case studies to be published in this series will be solicited from practitioners, researchers, and others working on the ground, who are affiliated with IPSI member organizations and closely involved and familiar with the activities related to SEPLS management.

Maiko Nishi • Suneetha M. Subramanian •
Philip Varghese
Editors

Business and Biodiversity

Reciprocal Connections in the Context of Socio-Ecological Production Landscapes and Seascapes (SEPLS)

Editors
Maiko Nishi
United Nations University Institute for the Advanced Study of Sustainability (UNU-IAS)
Tokyo, Japan

Suneetha M. Subramanian
United Nations University Institute for the Advanced Study of Sustainability (UNU-IAS)
Tokyo, Japan

Philip Varghese
Akita International University
Akita, Japan

United Nations University Institute for the Advanced Study of Sustainability (UNU-IAS)
Tokyo, Japan

This work was supported by the Ministry of the Environment, Japan (MOEJ)

ISSN 2731-5169 ISSN 2731-5177 (electronic)
Satoyama Initiative Thematic Review
ISBN 978-981-97-7573-6 ISBN 978-981-97-7574-3 (eBook)
https://doi.org/10.1007/978-981-97-7574-3

This Springer imprint is published by the registered company Springer Nature Singapore Pte Ltd.
The registered company address is: 152 Beach Road, #21-01/04 Gateway East, Singapore 189721, Singapore

Foreword

In confronting environmental and socio-political challenges, the critical role of businesses cannot be overemphasized. Regardless of size, location, or sector, every business relies on nature and its services, underpinned by biodiversity. According to the World Bank, a collapse of certain ecosystem services such as pollination and carbon sequestration could lead to losses of USD 2.7 trillion in the global economy by 2030, while low-income countries face an average 10% annual decline in GDP. Conversely, in 2020 the WEF emphasized that if companies add value to nature by making their activities nature-positive, they could embrace a business opportunity amounting to USD 10.1 trillion annually, creating 395 million jobs by 2030.

The importance of engaging businesses to attain the 2050 vision for biodiversity "living in harmony with nature" was recognized at the 15th Conference of the Parties to the Convention on Biological Diversity (CBD COP15) in December 2022. The Kunming–Montreal Global Biodiversity Framework, which was adopted at the conference, includes as one of its 23 global targets for 2030 facilitating nature-positive transitions in business and finance institutions (target 15). The United Nations University Institute for the Advanced Study of Sustainability closely supported the development of this new global framework through evidence-based contributions. We are committed to policy-relevant research, capacity development, and sustainability advocacy through mobilizing members of the International Partnership for the Satoyama Initiative (IPSI) across the globe.

In supporting implementation of the global biodiversity framework, in July 2023, IPSI adopted a new Strategy and Plan of Action for the period 2023–2030, which aligns its activities and approaches with those of the framework. One of its five strategic objectives focuses on sustainable value chain development. In this area, we are promoting nature-positive solutions that advance the IPSI vision of collaboration to realize societies in harmony with nature.

Facilitating nature-positive transitions in the business and finance community requires an explicit understanding of how businesses use natural resources and their diversity. It must also consider the various interactions between actors within the value chains through which the resources move. At the same time, we must ensure that policies and decisions fully embrace the principle of equity and leave no one behind. We are keenly aware that the voices of people involved in activities on the ground tend to be inadequately reflected in higher level decision-making. Often a lack of data and evidence on local actions and their impacts, challenges, and opportunities is cited as a constraint to inclusion. This publication series has been seeking to address precisely this challenge.

An important impact of the Satoyama Initiative Thematic Review (SITR) is its success in bringing local perspectives to the fore within policy discussions related to environment and development. Previous issues of the SITR series have explored the interlinkages between business and biodiversity—for instance, the connection between biodiversity mainstreaming and sustainable livelihoods. Building on these efforts, I am delighted that this volume is tackling the topic of engaging and strengthening business practices to further implementation of the new global biodiversity framework. The diverse case studies and synthesis of findings provide valuable insights and showcase many promising approaches. I sincerely hope that readers in all sectors will draw upon the contents of this publication to enhance and scale up efforts to address the biodiversity crisis and advance sustainable development.

United Nations University
Institute for the Advanced Study of
Sustainability
Tokyo, Japan

Shinobu Yume Yamaguchi

Foreword

Over the last decade, the United Nations University Institute for the Advanced Study of Sustainability (UNU-IAS) has been working with International Partnership for the Satoyama Initiative (IPSI) partners to curate experiences from diverse socio-ecological production landscapes and seascapes (SEPLS) that speak to important topics hotly discussed in policy forums such as the Convention on Biological Diversity (CBD) and the Intergovernmental Science-Policy Platform on Biodiversity and Ecosystem Services (IPBES), among others.

This volume focuses on the interlinkages between business and biodiversity, especially the dependencies of businesses on biodiversity and associated transactions and impacts, which is currently a topic of assessment that IPBES has initiated. We focus on the relevance of this topic in the context of SEPLS, to highlight local imperatives and peculiarities. In this volume too, I am pleased to note the diversity of experiences across the selected case studies. What is noteworthy is that despite the diversity, there are several common messages that include similar challenges and contextually driven approaches to arrive at outcomes that are supportive of local entrepreneurship, sustainable and equitable value chains, and flourishing ecosystems and biodiversity.

The IPSI has recently adopted its Strategy and Plan of Action with a renewed commitment to use knowledge products backed by evidence from the ground to inform policymaking on issues that promote social-ecological resilience. This means we will examine and highlight how integrated approaches on the ground can be operationalized in cooperation with IPSI partners. We aim to continue to provide credible, well-analysed information developed through a process of co-design and discursive analysis.

I hope you will find this volume engaging, and that the suggestions would strengthen the linkages between business practices and biodiversity conservation in a mutually reinforcing manner adaptable in your contexts and work.

Tsunao Watanabe

United Nations University
Institute for the Advanced Study of
Sustainability
Tokyo, Japan

The Secretariat of the International
Partnership for the Satoyama Initiative
Tokyo, Japan

Preface

The Satoyama Initiative is “a global effort to realize societies in harmony with nature”, started through a collaboration between the United Nations University (UNU) and the Ministry of the Environment of Japan. The initiative focuses on the revitalization and management of “socio-ecological production landscapes and seascapes” (SEPLS), areas where production activities help maintain biodiversity and ecosystem services in various forms while sustainably supporting the livelihoods and well-being of local communities. In 2010, the International Partnership for the Satoyama Initiative (IPSI) was established to implement the concept of the Satoyama Initiative and promote various activities by enhancing awareness and creating synergies among those working with SEPLS. IPSI provides a unique platform for organizations to exchange views and experiences and to find partners for collaboration. As of November 2024, 328 members have joined the partnership, including governmental, intergovernmental, non-governmental, private-sector, academic, and Indigenous Peoples’ organizations.

The Satoyama Initiative promotes the concept of SEPLS through a three-fold approach that argues for connection of land and seascapes holistically for management of SEPLS (see Fig. 1). This often means involvement of multiple sectors at the landscape or seascape scale, under which it seeks to: (1) consolidate wisdom in securing diverse ecosystem services and values, (2) integrate traditional ecosystem knowledge and modern science, and (3) explore new forms of co-management systems. Furthermore, activities for SEPLS management cover multiple dimensions, such as equity, addressing poverty and deforestation, and incorporation of traditional knowledge for sustainable management practices in primary production processes such as agriculture, fisheries, and forestry (UNU-IAS and IGES 2015).

As one of its core functions, IPSI serves as a knowledge-sharing platform through the collection and sharing of information and experiences on SEPLS, providing a place for discussion among members and beyond. Over 300 case studies have been collected and are shared on the IPSI website, providing a wide range of knowledge covering diverse issues related to SEPLS. Discussions have also been held to further strengthen IPSI’s knowledge-facilitation functions, with members suggesting that efforts should be made to produce knowledge on specific issues in SEPLS in order

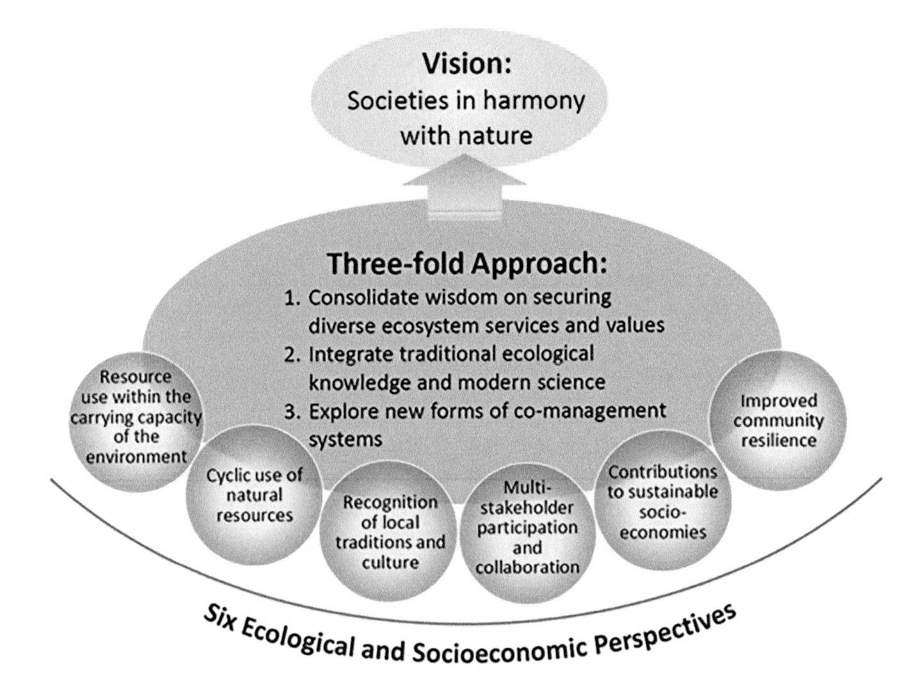

Fig. 1 The conceptual framework of the Satoyama Initiative

to make more targeted contributions to decision-makers and on-the-ground practitioners.

It is in this context that a project to create a publication series titled the "Satoyama Initiative Thematic Review" (SITR)[1] was initiated in 2015. The SITR series was developed as a compilation of case studies providing useful knowledge and lessons focusing on a specific theme that is important for SEPLS. The overall aim of the SITR publications is to collect experiences and relevant knowledge, especially from practitioners working on the ground, considering their usefulness in providing concrete and practical knowledge and information as well as their potential to contribute to policy recommendations. Each volume is also accompanied by a synthesis chapter which extracts lessons learned through the case studies, presenting them for policy-relevant academic discussions. This series also contributes to efforts being made by researchers to strengthen the evidence base for policymaking concerning social-ecological dynamics and resilience, including those under the Intergovernmental Science-Policy Platform on Biodiversity and Ecosystem Services (IPBES) and the Convention on Biological Diversity (CBD). Eight volumes have been published since 2015 for this series on various timely topics such as knowledge enhancement,

[1]The previous volumes of the SITR series are available at: https://satoyama-initiative.org/featured_activities/sitr/

biodiversity mainstreaming, sustainable livelihoods, effective area-based conservation, multiple values associated with sustainable use of biodiversity, transformative change, biodiversity-health-sustainability nexus, and ecosystem restoration.

Building on the earlier volumes that were in-house publications of UNU, the sixth edition and onward have been published by Springer Nature to reach out to a broader range of readers. In particular, the SITR has become a Springer book series from the seventh volume in the aim of ensuring consistent and coherent contributions to science-policy-practice interfaces, while enhancing the publications' impact and reach to a wide audience. Furthermore, a review committee has been established since the seventh volume by inviting experienced and knowledgeable experts in SEPLS management from the IPSI community to engage in the review process for publication. This expert group has helped to ensure credibility and reinforce the quality of the contents and at the same time to facilitate sharing of expertise and collaboration among IPSI members. The SITR series currently serves as one of the flagship projects underpinning the effort to attain "Knowledge Co-Production, Management, and Uptake", one of the five strategic objectives for the Strategy and Plan of Action 2023–2030 that was recently adopted at the Ninth IPSI Global Conference (IPSI-9) in 2023.

Similar to the previous volumes, this ninth edition was developed through a multi-stage process, including both peer review and discussions among the authors and reviewers at a workshop. Authors had several opportunities to receive feedback, which helped them to improve their manuscripts in substance, quality, and relevance. First, each manuscript received comments from the editors, the review committee members, and authors of two other chapters. Throughout the peer review process, each author received feedback from all these reviewers who were requested to comment on whether the manuscript was easy to understand and informative, addressed key questions of the volume's focus, and provided useful lessons as contributions to the theme of the volume. The aforementioned workshop was then held in person on 5–7 July 2023 to enable the exchange of feedback between authors and reviewers. The basic ideas contained in the synthesis of the concluding chapter were developed from presentations and discussions during the workshop, and the chapter was made available for two rounds of review by authors and reviewers before finalization.

The above process offers an opportunity for authors from both academic and non-academic organizations to contribute to generating knowledge in an accessible and interactive way, as well as to provide high-quality papers written in simple language for academics and a broader audience alike. It is our hope that this publication will be useful in providing information and insights to practitioners, researchers, and policymakers on the importance of long-term collaborative management of SEPLS that helps to promote and enhance sustainable business activities through which biodiversity can thrive. This, we hope, will prompt policymaking that strengthens such integrated and holistic management approaches.

We would like to thank all the authors who contributed their case studies. We also appreciate the continued commitment and support for the thorough review process by the four experts in the review committee: Yoji Natori, William Olupot, Anil

Kumar Nadesa Panicker, and Amanda Wheatley. We also thank the colleagues of UNU-IAS who were supportive and instrumental in organizing the workshop and facilitating the publication process: Tomoko Ago, Alexandra Franco Guajardo, Himangana Gupta, Maki Hosoda, Minako Ishizaki, Miles Lambert-Peck, Hideko Mimura, Rina Miyake, Makiko Mukaino, Nicholas Turner, Momoko Ueno, Makiko Yanagiya, Kanako Yoshino, and Madoka Yoshino. Our gratitude extends to the UNU administration, especially Francesco Foghetti, and to colleagues from United Nations Geospatial for all the instrumental support through the publication process. Furthermore, we acknowledge Susan Yoshimura who skilfully proofread the manuscripts.

Publication of this volume as a new Springer Book would not have been possible without the helpful guidance of Mei Hann Lee and Momoko Asawa from Springer and the institutional support and leadership of Shinobu Yume Yamaguchi and Tsunao Watanabe from UNU-IAS. Our grateful thanks are also due to the Ministry of the Environment, Japan for supporting the activities of IPSI and its secretariat hosted by UNU-IAS.

Tokyo, Japan — Maiko Nishi
Tokyo, Japan — Suneetha M. Subramanian
Akita, Japan — Philip Varghese

Reference

UNU-IAS & IGES (eds.) (2015) Enhancing knowledge for better management of socio-ecological production landscapes and seascapes (SEPLS). Satoyama Initiative Thematic Review, vol. 1, United Nations University Institute for the Advanced Study of Sustainability, Tokyo

Acknowledgements

Review Committee

Yoji Natori
William Olupot
Anil Kumar Nadesa Panicker
Amanda Wheatley

Editorial Support

Madoka Yoshino

English Proofreading

Susan Yoshimura

Contents

Editors and Contributors

About the Editors

Maiko Nishi Research Fellow at United Nations University Institute for the Advanced Study of Sustainability. Her research interests include social-ecological system governance, regional planning, and agricultural land policy. PhD in Urban Planning from Columbia University.

Suneetha M. Subramanian Research Fellow at United Nations University Institute for the Advanced Study of Sustainability. Her research interests include biodiversity and human well-being with a focus on equity, traditional knowledge, community well-being, and social-ecological resilience.

Philip Varghese JSPS-UNU Postdoctoral Fellow at Akita International University, Japan and United Nations University Institute for the Advanced Study of Sustainability. His research interests include tourism and development politics, indigenous communities, and sustainability.

Contributors

William Dunbar Project Manager at Conservation International in Tokyo, Japan, focusing on international sustainable landscape and seascape policies, particularly relating to landscapes as "other effective area-based conservation measures" (OECMs).

Lorena Jaramillo Economic Affairs Officer and the project manager of the Global BioTrade Programme at UNCTAD Switzerland, working in areas of trade and biodiversity, sustainable trade, and business development.

Lika Sasaki Programme Management Officer at UNCTAD Switzerland for the Global BioTrade Programme. BioTrade promotes the sustainable trade of biodiversity-based products and services in a way that can benefit nature and sustain livelihoods.

Emilio R. Díaz-Varela Associate Professor at the University of Santiago de Compostela, Spain. His research interests are social-ecological systems analysis and management, spatial analysis in complex patterns, and sustainable development and resilience in urban and rural systems.

Pedro Álvarez-Álvarez Associate Professor in the Department of Organisms and Systems Biology, University of Oviedo, Spain. Pedro is interested in sustainable forest management, complex forests, and traditional chestnut (Castanea sativa Mill.) stands.

José C. Pérez-Girón Postdoctoral researcher at the Department of Botany of the University of Granada, Spain. He is interested in primary production indicators and the potential effects of climate change on plant and animal species.

Ramón A. Díaz-Varela Associate Professor at the University of Santiago de Compostela, Spain. Research interests are the use of geomatics for the assessment of biodiversity at different scales.

María A. Ferreira Golpe Researcher in the Department of Applied Economics at the University of Santiago de Compostela, Spain. Works on projects focused on beekeeping, invasive species, and rural policy.

Ana I. García-Arias Associate Professor in the Department of Applied Economics at the University of Santiago de Compostela, Spain. Her publications are related to agricultural economics, the evaluation of agri-environmental policies, and the identification of ecosystem services.

Guido Gualandi Teaches History of Food in the Mediterranean at Gonzaga University in Florence and Accent International. He has worked as an archaeologist in France, Italy, and the Middle East. He owns a farm in Tuscany and is a member of the Ancient Grains Community in Montespertoli, Italy.

Dambar Pun CEO of Back to Nature, Nepal. Has worked on conservation and ecotourism-based business since 2003. Has conducted philanthropic work around the Panchase Protected Forest for community awareness and skill development on biodiversity conservation and ecotourism since 2011.

Aashish Tiwari Works at the Conservation Development Foundation, Nepal. Enthusiastic on conservation-based natural resource management. Has a Master's degree in forest management and biodiversity conservation from the Institute of Forestry, Pokhara, Nepal.

Rebecca Gurung Sustainability Officer involved with the Himalayan Sustainable Future Foundation. Has a Master's in environment management from Lincoln University, New Zealand. She has been working in the field of biodiversity conservation and climate change in Western and Mid-Western Nepal.

Hum Bahadur Gurung Asia Partnership Manager at Bird Conservational Nepal (BirdLife International in Nepal). He has a PhD from Griffith University, Australia. His research interests include participatory action and community-based approaches for environmental education, sustainable tourism management, and ecosystem services.

Samuel Pun Serves as Managing Director at Backyard by Back to Nature, a restaurant specializing in promoting eco-friendly and sustainable living practices, while studying at Westcliff University. Dedicated to sustainability, he contributes to preserving the environment and promoting a greener lifestyle.

K. G. Saxena Currently an independent researcher in inter-disciplinary and multi-institutional endeavors for the promotion of sustainable farms, forests, landscapes, and livelihoods. Former Dean (Environmental Sciences) and Professor (Ecology and Sustainable Development) of Jawaharlal Nehru University, New Delhi, India.

S. Sreekesh Professor (Geography) at Jawaharlal Nehru University, New Delhi, India, leading programmes on climate change-induced physical and human vulnerabilities and adaptations.

K. S. Rao Senior Professor in the Department of Botany, University of Delhi, India. Currently involved in academic work related to natural resource management and biodiversity assessment in critical ecosystems in hill and mountain regions.

R. K. Maikhuri Professor in the Department of Environmental Sciences, HNB Garhwal University, Srinagar, Uttarakhand, India, with over 38 years of research/development experience in the interphase area of ecology-natural resource management and sustainable development in the Indian Himalayan region.

S. Nautiyal Director, G.B. Pant National Institute of Himalayan Environment (Ministry of Environment, Forests and Climate Change, Government of India), Almora, and Professor, Institute for Social and Economic Change, Bengaluru, India, leading initiatives on natural resource management and socio-economic and ecological approaches to sustainable development.

Jude C. Baggo Director for the Globally Important Agricultural Heritage Systems Center of the Ifugao State University. The Center focuses on human capacity building towards the Ifugao Rice Terraces and its attributes.

Eva Marie Codamon-Dugyon President of Ifugao State University. She champions the conservation of heritage, biodiversity, and sustainable development.

Clyde B. Pumihic International Relations Officer of Ifugao State University. His research interests include indigenous cuisines and heritage conservation.

Marah Joy A. Nanglegan Director for the Department of Extension and Training of Ifugao State University. She is an expert in development communication.

Florence Mayocyoc-Daguitan Coordinator of the Indigenous Peoples and Biodiversity Program at Tebtebba, Indigenous Peoples' International Centre for Policy Research and Education in the Philippines.

Guesang Farmers' Organization Inc A multi-sectoral community organization composed mainly of farmers based in Guesang, Bangaan, Sagada, Mountain Province in the Philippines.

Shyh-Huei Hwang Distinguished Professor and Library Director at National Yunlin University of Science and Technology (YunTech), Chinese Taipei.

Hsiu-Mei Huang Doctoral student at Graduate School of Design, National Yunlin University of Science and Technology, Chinese Taipei.

Tzu-Hsuan Chan Has a Master's degree from the Graduate Institute of Architecture and Cultural Heritage, Taipei National University of the Arts.

Mbole Véronique Project Assistant at Green Development Advocates, Cameroon, working on projects focused on assessing the policies and impacts of large infrastructural projects on surrounding communities and assisting communities in agro-ecological and seed conservation practices.

Aristide Chacgom Fokam Coordinator and Environmental Jurist at Green Development Advocates, Cameroon, working on projects focused on assessing the impacts of policies on different actors, particularly communities around large infrastructural and agro-industrial projects and protected areas.

Alain Fabrice Mfoulou Bonny PhD candidate in public law and a Jurist at Green Development Advocates, Cameroon. Works on projects focused on assessing the impacts of policies on different actors, particularly communities around large infrastructural and agro-industrial projects.

Carrele Floriane Nguena Mawamba Project Officer at Green Development Advocates, Cameroon. Works on projects focused on mapping out community spaces and assessing the impacts of agro-industrial projects on surrounding forest communities and assisting communities in agro-ecological and seed conservation practices.

Stéphane Nzakou Tchakounte Project Assistant at Green Development Advocates, Cameroon. Works on projects focused on mapping out community spaces and assessing the impacts of agro-industrial projects on surrounding forest communities and assisting communities in agro-ecological and seed conservation practices.

Laetitia Musi Adjoffon Research Intern at Green Development Advocates, Cameroon. Works on projects focused on evaluating how ecosystem services affect and are affected by modified land use practices in the transformation of forest-agriculture boundaries.

Corine Linda Ehowe Issova Research Intern at Green Development Advocates, Cameroon. Works on projects focused on evaluating the well-being of forest communities around agro-industrial and large infrastructural projects.

Andrés Quintero-Ángel PhD candidate in environmental sciences and Scientific and Research Director of Corporación Ambiental y Forestal del Pacifico (CORFOPAL), Colombia. He majored in conservation and use of biodiversity with ethnic communities.

Sebastian Orjuela-Salazar Specialist in planning, follow-up and monitoring of the Fund for Environmental Action and Children - Fondo Acción.

Diana Saavedra-Zuñiga Social Worker specializing in social management, with 14 years of experience as a consultant and coordinator of disaster risk management projects, natural protected areas, and watershed management plans.

Alejandro Castaño-Astudillo Biologist specializing in environmental education, with experience in accompaniment processes and community capacity building for environmental management, nature tourism, governance, and environmental education.

María Viviana Borda Calvache Researcher at the Research group on sustainable socio-ecological systems (SSES). She has a Master's degree in conservation and use of biodiversity.

Mauricio Quintero-Ángel Has a PhD in environmental sciences and works as an Associate Professor at Universidad del Valle, Colombia. Interested in research on social-ecological systems, landscape planning, and rural development.

Guanqi Li Director of East Office of Farmers' Seed Network (China). Works on projects focused on farmers' seed systems enhancement to build practice-science-policy linkage and promote agricultural biodiversity conservation and utilization that enhances food security and sustainable food systems.

Xin Song Director of Southwest Office of Farmers' Seed Network (China). Focuses on community-based participatory plant breeding and related action research to improve small farmers' resilience in Southwest China.

Ye Shen Coordinator at Farmers' Seed Network (China). Work on projects focused on seed conservation and utilization in the context of climate change to build "From Seed to Table" initiative that promotes sustainable seed and food systems.

Uma Khumairoh Researcher and Lecturer in the Integrated Organic Farming Systems Research Centre, Faculty of Agriculture, Brawijaya University, and a guest researcher and lecturer at Wageningen University. She specializes in agroecology, complex agriculture design, biocultural farming, and farmer community development.

Rochmatin Agustina Researcher and Lecturer in the Agricultural Faculty, University of Muhammadiyah Gresik. Her specialization is in organic plant production.

Euis Elih Nurlaelih Researcher and Lecturer in the Integrated Organic Farming Systems Research Centre, Faculty of Agriculture, University of Brawijaya. She specializes in Indonesian traditional home gardens and ecological design of agriculture.

Dewi Ratih Rizki Damaiyanti Researcher and Lecturer in the Integrated Organic Farming Systems Research Centre, Faculty of Agriculture, University of Brawijaya. Her specialization is in plant ecology and plant production.

Adi Setiawan Researcher and Lecturer in the Integrated Organic Farming Systems Research Centre, Faculty of Agriculture, University of Brawijaya. He specializes in tree cultivation and plant ecology.

Jeroen C. J. Groot Associate Professor at the Farming Systems Ecology group of Wageningen University, the Netherlands. He specializes in farming systems analysis and agroecology.

Chapter 1
Introduction

Maiko Nishi, Suneetha M. Subramanian, and Philip Varghese

Abstract This chapter lays out a context for discussing the interconnections between business and biodiversity in light of the management of socio-ecological production landscapes and seascapes (SEPLS). First, it provides a general picture of how businesses both impact and depend on biodiversity. It also introduces recent initiatives and associated challenges in making businesses more sustainable and having businesses and industries engage in biodiversity conservation and sustainable use. Second, the chapter discusses the relevance of SEPLS to businesses. In particular, it highlights potential implications that lessons learned from case studies on SEPLS management have for innovative solutions to facilitate more sustainable business decisions and actions. Finally, it outlines the scope, objectives, and organization of the book, including an overview of the case studies compiled in the subsequent chapters.

Keywords Socio-ecological production landscapes and seascapes · Businesses · Biodiversity · Sustainability · Value chains · Landscape approaches · Case studies · Science-policy-practice interface

M. Nishi (✉) · S. M. Subramanian
United Nations University Institute for the Advanced Study of Sustainability (UNU-IAS), Tokyo, Japan
e-mail: nishi@unu.edu

P. Varghese
United Nations University Institute for the Advanced Study of Sustainability (UNU-IAS), Tokyo, Japan

Akita International University, Akita, Japan

M. Nishi et al. (eds.), *Business and Biodiversity*, Satoyama Initiative Thematic Review, https://doi.org/10.1007/978-981-97-7574-3_1

1 Business and Biodiversity

Businesses both impact and depend on biodiversity, not only because of the immediate use of natural resources and its direct effects, but also indirectly through supply chains, value chains,[1] regulatory systems, financing mechanisms, and consumer reputations (WEF 2010). Despite the varied degree of interconnectivities, almost all businesses essentially rely on biodiversity—ranging from small- and mid-sized enterprises to globally operating companies across different sectors (e.g. primary industries, food processing, pharmaceutical and healthcare, tourism, financial, utilities) (Smith et al. 2020). Along with the declining trend in biodiversity, business risks associated with biodiversity loss have been increasingly recognized worldwide. The fifth edition of the Global Risks Report released by the World Economic Forum (WEF) in 2010 featured, for the first time, "biodiversity loss" as one of the major global risks with the potential of systemic failures (Dempsey 2013). The current trend is even more worrisome, as USD 44 trillion of economic value generation (i.e. more than half of the global GDP) appears to be moderately or highly dependent on nature and is thus vulnerable to biodiversity loss and ecosystem degradation (WEF 2020a).

While affirming "no sector escapes untouched by some form of biodiversity risk," the WEF also highlights opportunities, for instance, with new technologies, innovative trading systems, improved models for land and sea uses, and novel market mechanisms (WEF 2010, p. 8). Positioning businesses as major drivers of ecosystem change points to their potential role in radically reducing their negative impacts (IPBES 2019). When businesses prevent, decrease, mitigate, or manage risks associated with the biodiversity they essentially rely on, they have opportunities to curtail or mitigate their negative impacts on biodiversity through reducing pressure on nature from resource extraction or waste production (e.g. circular economy, waste prevention, pollution control) (TNFD 2023a).

Furthermore, proactive engagement by businesses in making socially and ecologically sound impacts (e.g. conservation, restoration, and sustainable use of biodiversity, or financing or insurance to support them) may create new values, putting in place sustainable business models (e.g. ethical sourcing, social enterprise, demand management) (Bocken et al. 2014; TNFD 2023a). This can bring in profit

[1] The term of value chain refers to the functional business activities wherein each stage of the chain adds more value towards its customers (Dubey et al. 2020). For instance, a medicinal plant converted to dry powder can be further converted to a botanical supplement. A supply chain comprises the steps it takes to deliver the product or service from its original state to the customer, consisting of all contributors involved in the development of a product, from sourcing of raw materials to final distribution of a finished product to customers (Dubey et al. 2020). An example of a supply chain is a medicinal plant harvested by local communities and then assembled by local agents being sold to wholesalers and retailers, who then sell to intermediary processors who convert plants to powders and sell to the end user to make botanical supplements who finally market it to end-user consumers. The major distinctions between these concepts are their focuses and directions: a supply chain is generally directed towards the supply base, while a value chain is geared to the customer (Feller et al. 2006).

by seizing new business opportunities, strategically changing products and services and investments, creating new markets, and developing additional revenue streams (TNFD 2023a). It has been estimated that these opportunities can add up to USD 10.1 trillion in annual business value and create 395 million jobs by 2030 through nature-positive solutions (WEF 2020b). Given their critical role in determining societal impacts on biodiversity, businesses have great potential to lead transformative change by facilitating systemic improvements that speak to multiple leverage points (e.g. total consumption and waste, values and action, externalities and telecouplings, and innovation and investment) (IPBES 2019; zu Ermgassen et al. 2022).

With the growing recognition of their key role in systemic transformations, businesses have become increasingly committed to and involved in the processes to curve the trends of biodiversity loss and create positive effects on nature (de Silva et al. 2019). New tools and regulations are also emerging to help businesses make informed decisions and navigate appropriate actions for biodiversity conservation (Stephenson and Walls 2022). For instance, officially established in June 2021, the Taskforce on Nature-Related Financial Disclosures (TNFD) Framework was endorsed by the G7 Finance Ministers, and recommendations and additional guidance were published in 2023 to assist companies in collating and reporting their nature-related risks and opportunities (White et al. 2023; TNFD 2023b). In addition, various business-led platforms have been developed to help measure business impacts on biodiversity and develop strategies and plans for biodiversity conservation—including Finance for Biodiversity, Business for Nature, Act4Nature, and the Science Based Targets Network (SBTN) to name a few (White et al. 2023).

However, business engagement in countering biodiversity loss remains limited, making only marginal or sporadic improvements (Smith et al. 2020; TNFD 2023a). Besides the lack of obligations or incentives from governments and market regulators, challenges exist largely in understanding, measuring, and communicating the interdependency between businesses and biodiversity (TNFD 2023c). In particular, the complexity of the connections, where many ecosystem services are invisible, makes it difficult for companies to grasp them comprehensively (Dasgupta 2021). Even in primary industry sectors, where the connections to biodiversity are rather obvious, links could be blurred when multiple product lines are developed alongside long supply chains (White et al. 2023). For instance, the fashion industry depends on land, plants, and freshwater to produce raw fibre materials, while causing environmental externalities (e.g. pollution) across its value chain. Such impacts and dependencies are not easily apprehended, challenging informed decision-making (Stephenson and Walls 2022). In this context, the Intergovernmental Science-Policy Platform on Biodiversity and Ecosystem Services (IPBES) is currently undertaking a methodological assessment of the impact and dependence of business on biodiversity and nature's contributions to people (the so-called "business and biodiversity assessment"). It aims to categorize how businesses depend on, and impact, biodiversity and nature's contributions to people and identify criteria and indicators for measuring that dependence and impact.[2]

[2] https://ipbes.net/business-impact

2 Relevance of Socio-Ecological Production Landscapes and Seascapes to Sustainable Business Model

People active in socio-ecological production landscapes and seascapes (SEPLS) derive different types of benefits from these spaces. A diversity of resources and a mosaic of ecosystems (combination of different types of ecosystems) in these areas are used for livelihood and various types of production and value addition activities, and cultural, recreational, and educational purposes among others. Based on management practices that balance conservation and development needs, local actors steward land and sea and maintain biodiversity, while integrating traditional ecological knowledge and modern science to promote innovations and value additions that contribute to the society in many ways. What is noteworthy is that these actors recognize the interdependence between ecological and social resilience and factor it into decisions related to use of the ecosystems and related resources.

That said, such sites are also prospected for business activities with varying consequences. These could be positive when the impacts on both the environment and social transactions are aligned with principles of sustainable use and equity. However, these could turn out to be adverse when benefits are unevenly distributed, and the land/seascape becomes degraded by losing its multifunctional nature and social-ecological integrity due to various reasons such as unsustainable production practices and large-scale infrastructure development. Such actions are often driven by policy decisions, support for new economic opportunities, and changes to demography among others. Furthermore, it is possible that these spaces are affected due to activities that happen in distant locations, and that are connected to resource use or human endeavours on these sites.

Despite intricate interactions of these actions and effects, local actors in SEPLS, who often consist of the upstream end of long and complex value chains, observe and experience profound impacts of businesses within their intimate environment, including both positive (e.g. employment, branding of the place) and negative (e.g. ecosystem degradation, deprived rights to territories) impacts. Their first-hand knowledge obtained through direct observation and experiences should inform the decisions and actions of multiple actors—including not only businesses but also those who operate, regulate, and are affected by and concerned with business activities. In this regard, case studies on SEPLS management from the local perspective provide rich evidence to detail and help clarify the interconnections between businesses and biodiversity, while exemplifying innovative solutions that combine different types of expertise (modern and traditional) to facilitate more sustainable business decisions and actions for socially and ecologically sound outcomes. Furthermore, the experiences in managing SEPLS where multiple actors negotiate and collaborate for meeting diverse needs and interests provide practical insights on the roles and responsibilities of stakeholders as well as how to build partnerships to promote systemic changes towards sustainable futures.

3 Objectives and Structure of the Book

In this volume, we seek to enhance our understanding of the implications of business decisions and practices on SEPLS and conversely, on the resourcefulness of SEPLS on business activities. This we do through evidence collected from the partners committed to the International Partnership for the Satoyama Initiative (IPSI). With empirical evidence provided in the case studies (Chaps. 2, 3, 4, 5, 6, 7, 8, 9, 10, 11, 12, and 13) and the synthesis of key findings derived from these studies (Chap. 14), we aim to highlight how businesses are dependent on and impact biodiversity and natural environments in SEPLS, while addressing how SEPLS management practices support business activities that rely on natural resources and render positive biodiversity outcomes.

The case studies compiled in this volume demonstrate experiences and insights on the roles, attitudes, motivations, and actions of multiple stakeholders, on-the-ground impacts of businesses in SEPLS, and methodologies to measure impacts and dependency of business on biodiversity and nature's contributions to people, among others. They broadly address the following questions:

- What and how multiple benefits derived through SEPLS management have directly or indirectly contributed to any business activity primarily relying on biodiversity and natural resources?
- Are there any (positive and/or negative) impacts on SEPLS made by business activities (or certain market mechanisms)? If so, what are they, and who has been affected? How can you measure and assess these impacts?
- What efforts have been made to address (positive and/or negative) impacts of businesses in managing SEPLS—including the development or adoption of sustainable development models? What factors have influenced decisions and actions in dealing with business impacts on SEPLS? How have different stakeholders negotiated and collaborated to manage the business impacts on SEPLS?
- Has local and traditional knowledge and cultural diversity helped to inform or facilitate business decisions and actions for socially and ecologically sound outcomes? If so, how?

As mentioned above, the subsequent Chaps. 2, 3, 4, 5, 6, 7, 8, 9, 10, 11, 12, and 13 exemplify 12 case studies from different parts of the world, comprising two from Europe, one from Latin America, one from Africa, and eight from Asia (Fig. 1.1). All the cases are grouped under five different thematic areas on the basis of the ecosystem types and the nature of major products: (1) mountain and forest products, (2) forest products, (3) mountain and forest-based industry, (4) mountain, forest and agricultural products, and (5) agricultural products (Table 1.1).

Each case study portrays a distinct story focused on several rice-based products (Chaps. 7, 8, 12, and 13), non-rice products (Chaps. 2, 3, 4, 10, and 11), and ecotourism experiences and cultural practices (Chaps. 5, 6, and 9) and their dependencies and impacts on biodiversity and SEPLS. The cross-cutting connections between

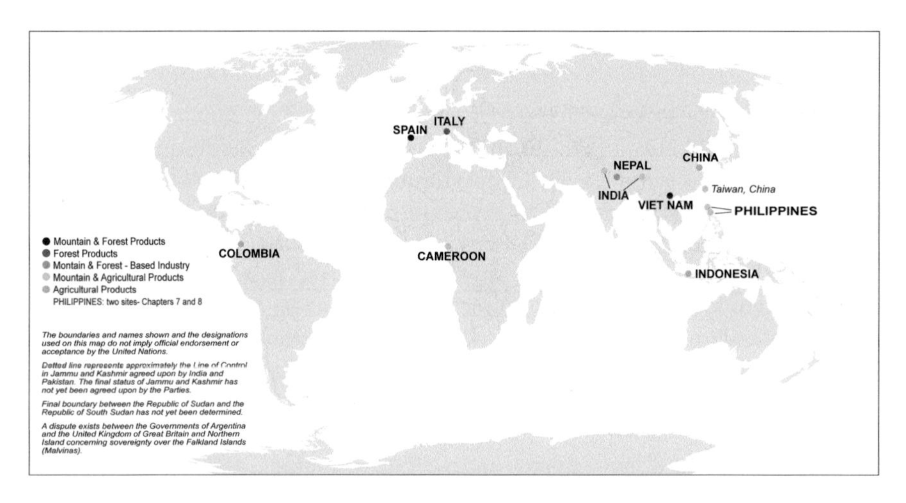

Fig. 1.1 Locations of the case studies (map template: United Nations Geospatial). Note: Details of the case study locations, including geographic coordinates, are described in each chapter

them are the challenges of natural resource exploitation, habitat loss, local well-being, lower yields, economic inflation and leakage, loss of interest among younger people, Indigenous practices, cultural diversity and values, livelihoods of local communities, soil fertility, and land use change. Notably, we also see individual and collective efforts—both existing and proposed initiatives at local, national, regional, and international levels—in ethical standards in business, fair and equitable sharing of benefits, conservation of biodiversity, value addition, nature-positive approach, scientific improvements and innovations, social enterprises, stakeholder mobilization, community-led sustainable agricultural practices, better monitoring and evaluation practices, and promotion of ecotourism to deliver long-term solutions to address business-induced stress and trade-offs.

Lastly, Chap. 14 is the synthesis chapter, which is a summation of key messages from each of the case studies. Irrespective of the scale of business operations, market reach and dependency, and impact on the ecosystem services and biodiversity, all case studies put forward valuable insights regarding the reciprocal connections of business and biodiversity in the context of SEPLS. The synthesis aims to draw the conceptual connection between business and biodiversity and its prognosis, offer methodologies for monitoring and evaluation, and ways to address challenges and seize opportunities in SEPLS to effectively manage business impacts on biodiversity, ecosystems, and human well-being. Further, this chapter outlines recommendations for policymakers and other stakeholders to ensure continued multiple benefits from SEPLS through ecologically and socially sound and equitable processes while promoting sustainable businesses.

A summary overview of the case studies is given in Table 1.1, while the map indicates the locations of the case studies (Fig. 1.1).

Table 1.1 Overview of the case studies

Focused thematic areas	Chapter (Country)	Ecosystem types	Business types	Business dependencies and impacts on biodiversity and SEPLS	Efforts (including those proposed) to address business impacts
Mountain and forest products	Chap. 2 (Viet Nam)	Mountain, forest, grassland	Siam benzoin gum production and associated business and trade	• Dependencies: *Styrax tonkinensis* tree forest for Siam benzoin gum production • Impacts: Change in ecosystems, ecological processes, natural habitats, and species, particularly threatened or endangered species	BioTrade Principles and Criteria (P&C) support conservation of biodiversity, fair and equitable sharing of benefits, and respect for the actors involved; improving local well-being through sustainable trade
	Chap. 3 (Spain)	Mountain, forest, agricultural	Chestnut production and associated businesses (e.g. chestnut flour, timber, mushrooms)	• Dependencies: Well-maintained chestnut forests • Impacts: Poor maintenance leading to lower yields and rise in wildfires	Value added products, marketing to external markets, innovatively supporting production, trade, and promotion of local chestnut varieties through protected geographical indication; raising cultural sense of place through education and locally important activities
Forest products	Chap. 4 (Italy)	Forest, agricultural, coastal	White truffle harvest and associated businesses (tourism, food, etc.)	• Dependencies: Harvest of wild white truffles (declining due to climate change, biodiversity loss, inappropriate soil and water management, industrial agricultural practices in surrounding areas, and loss of traditional environmental knowledge) • Impacts: Increased tourism and hospitality, local economy, food culture and other cultural activities associated with truffles	Fair and equitable benefit-sharing among landowners, farmers (primary land users), harvesters, and other value-chain actors; government interventions to facilitate land-use for truffle conservation

(continued)

Table 1.1 (continued)

Focused thematic areas	Chapter (Country)	Ecosystem types	Business types	Business dependencies and impacts on biodiversity and SEPLS	Efforts (including those proposed) to address business impacts
Mountain and forest-based industry	Chap. 5 (Nepal)	Mountain, forest, grassland, agricultural, inland water	Ecotourism	• Dependencies: Natural environment, resources, and critical habitat • Impacts: Biodiversity loss, pollution, climate change, environmental social degradation, economic inflation and leakage, loss of habitats and wildlife species, and various other social issues	Nature-positive approach for ecosystem restoration and preservation and protection of critical habitat; farm to table experience, employment opportunities, waste management and conservation initiatives through sustainable ecotourism business model
Mountain, forest and agricultural products	Chap. 6 (India)	Forest, grassland, agricultural, dryland	Agroforestry	• Dependencies: Tropical, temperate, and alpine biomes • Impacts: Livelihood, cultural values, traditional practices, policy-led changes, subsidies, herbal medicines, market of organic food products and other non-timber forest products	Scientific improvements in traditional manure production, agroforestry, adoption of vermitechnology, rehabilitation of degraded lands, and people-led ecotourism; cooperative marketing/value addition, business/market risks and uncertainties, and participatory research are to be included in community forestry
	Chap. 7 (Philippines)	Mountain, forest, grassland, (agricultural)	Production and marketing of rice-based tea and taro-based ice cream	• Dependencies: Ifugao Rice Terraces (threatened due to climate change and outmigration) • Impacts: Ifugao satoyama, livelihood, ageing population, Indigenous practices and loss of interest among young people	Regular training programmes for revitalization of the rice terraces through product innovation, use of Indigenous practices, and promotion of locally owned small businesses
	Chap. 8 (Philippines)	Mountain, forest, grassland, agricultural, inland water	Rice production, poultry, animal raising	• Dependencies: Land, water, soil, and animals • Impacts: Better soil quality, consequently better-quality crop yield, higher control of community over economy	Social enterprise focusing on protecting traditional territories, culture; expanding to tourism, weaving, food processing, and artisanal mining, organic farm inputs
	Chap. 9 (Chinese Taipei)	Mountain, forest, agricultural	Weaving ramie, cultural businesses	• Dependencies: Special reed needed for fibre and supporting ecosystem • Impacts: Reducing number of practitioners	Value addition, linking to modern markets, and partnerships with university for technical input

Agricultural products	Chap. 10 (Cameroon)	Mountain, forest, agricultural, coastal	Palm plantation, oil processing	• Dependencies: Land (forest, soil) and water • Impacts: Threats to biodiversity and livelihoods of local communities	Advocacy campaigns to defend local communities' rights and livelihoods; stakeholder mobilization (communities, civil society, media, women); support for community-led agro-ecological practices
	Chap. 11 (Colombia)	Forest, agriculture, inland water	Brown sugar cane production and processing	• Dependencies: Tropical dry and very dry (subxerophytic) forests and their productivity for sugar cane production (water regulation and soil formation) • Impacts: External pressures on the forests (land-use change, economic development)	Establishment of a protected area; development of sustainable model practices by community-based association of sugar cane mill enterprises
	Chap. 12 (China)	Agricultural, inland water, peri-urban	Organic rice farming	• Dependencies: Rice germplasm diversity and ecological integrity to support rice production ecosystem • Impacts: Better options, especially for climate adapted varieties	Using diversity to promote ecotourism, developing high value rice, organic farming, value-added products such as specialty rice wine, and improving crop resilience
	Chap. 13 (Indonesia)	Agricultural	Rice production	• Dependencies: Rice agroecosystems for the production of rice and other crops • Impacts: Environmental and cultural diversity, agricultural income	Farmer field schools (FFS) with demonstration farms on complex rice systems (CRS); cultivation of multiple species on rice bunds

References

Bocken NMP, Short SW, Rana P, Evans S (2014) A literature and practice review to develop sustainable business model archetypes. J Clean Prod 65:42–56

Dasgupta P (2021) The economics of biodiversity: the Dasgupta review. Hm Treasury, London

de Silva GC, Regan EC, Pollard EHB, Addison PFE (2019) The evolution of corporate no net loss and net positive impact biodiversity commitments: Understanding appetite and addressing challenges. Bus Strategy Environ 28(7):1481–1495

Dempsey J (2013) Biodiversity loss as material risk: Tracking the changing meanings and materialities of biodiversity conservation. Geoforum 45:41–51

Dubey S, Singh R, Singh SP, Mishra A, Singh NV (2020) A brief study of value chain and supply chain. In Rao RK (ed) Agriculture Development and Economic Transformation in Global Scenario, Mahima Research Foundation and Social Welfare, India, pp. 177–183

Feller A, Shunk D, Callarman T (2006) Value chains versus supply chains. BP Trends 1:1–7.

IPBES (2019) Global assessment report of the Intergovernmental Science-Policy Platform on Biodiversity and Ecosystem Services, Brondízio ES, Settele J, Díaz S, Ngo HT (eds). IPBES secretariat, Bonn, Germany. 1144 pages. ISBN: 978-3-947851-20-1

Smith T, Beagley L, Bull J, Milner-Gulland EJ, Smith M, Vorhies F, Addison PF (2020) Biodiversity means business: Reframing global biodiversity goals for the private sector. Conservation Letters 13(1):e12690

Stephenson PJ, Walls J (2022) A new biodiversity paradigm for business. Amplify 35:6–14

TNFD (2023a) Guidance on the identification and assessment of nature-related issues: The LEAP approach. Version 1.1, October 2023. https://tnfd.global/publication/additional-guidance-on-assessment-of-nature-related-issues-the-leap-approach/

TNFD (2023b) Recommendations of the Taskforce on Nature-related Financial Disclosures. September 2023. https://tnfd.global/wp-content/uploads/2023/08/Recommendations_of_the_Taskforce_on_Nature-related_Financial_Disclosures_September_2023.pdf?v=1695118661

TNFD (2023c) Getting started with adoption of the TNFD recommendations. Version 1.0 September 2023. https://tnfd.global/wp-content/uploads/2023/09/Getting_started_TNFD_v1.pdf?v=1698156380

WEF (2010) Biodiversity and Business Risk. WEF, Geneva

WEF (2020a) Nature risk rising: Why the crisis engulfing nature matters for business and the economy. In collaboration with PwC. New Nature Economy series

WEF (2020b) New Nature Economy Report II—The Future of Nature and Business. WEF, Cologne

White TB, Petrovan SO, Bennun LA, Butterworth T, Christie AP, Downey H, Hunter SB, Jobson BR, zu Ermgassen SOSE, Sutherland WJ (2023) Principles for using evidence to improve biodiversity impact mitigation by business. Bus Strateg Environ 32(7):4719–4733. https://doi.org/10.1002/bse.3389

zu Ermgassen SO, Howard M, Bennun L, Addison PF, Bull JW, Loveridge R, Pollard E, Starkey M (2022) Are corporate biodiversity commitments consistent with delivering ‘nature-positive’ outcomes? A review of ‘nature-positive’ definitions, company progress and challenges. J Clean Prod 379:134798

Maiko Nishi Research Fellow at United Nations University Institute for the Advanced Study of Sustainability. Her research interests include social-ecological system governance, local and regional planning, and agricultural land policy.

Suneetha M. Subramanian Research Fellow at United Nations University Institute for the Advanced Study of Sustainability. Her research interests include biodiversity and human well-being with a focus on equity, traditional knowledge, community well-being, and social-ecological resilience.

Philip Varghese JSPS-UNU Postdoctoral Fellow at Akita International University, Japan and United Nations University Institute for the Advanced Study of Sustainability. His research interests include tourism and development politics, indigenous communities, and sustainability.

Chapter 2
BioTrade Production and Sourcing of Siam Benzoin Gum in Northern Viet Nam

William Dunbar, Lorena Jaramillo, and Lika Sasaki

Abstract This chapter presents a case study carried out under BioTrade, an initiative headed by UNCTAD to promote sustainable and biodiversity-friendly trade. The case is related to production of Siam benzoin gum in Viet Nam supported under the Helvetas' Regional BioTrade Project, and demonstrates well how sustainable trade can lead to effective biodiversity conservation and livelihood benefits for those responsible for managing needed resources. Impacts of the programme are assessed through internal evaluations and published literature. BioTrade is an example of what can happen when a product or service sourced from biodiversity is commercialized and traded in a way that respects people and nature, as in the case of benzoin gum. BioTrade has established a set of Principles and Criteria to support the conservation and sustainable use of biodiversity, as well as the fair and equitable sharing of benefits through trade, and these are applied on the ground by businesses and other organizations around the world, including the highlighted case in Viet Nam. Future work is expected to expand on these successes and also to strengthen the SEPLS-management aspect of the work through stronger cooperation with the Satoyama Initiative.

Keywords Biodiversity · Trade · BioTrade · Siam benzoin gum · Viet Nam · Deforestation · Community benefits

1 Introduction

This chapter presents a case study of a project promoting the sustainable production and trade of Siam benzoin gum products from Viet Nam, supported by Helvetas' Regional BioTrade Project (Helvetas 2023) under the umbrella of the BioTrade Initiative, a programme spearheaded by the United Nations Conference on Trade

W. Dunbar (✉)
Conservation International Japan, Shinjuku, Tokyo, Japan

L. Jaramillo · L. Sasaki
United Nations Conference on Trade and Development (UNCTAD), Geneva, Switzerland

M. Nishi et al. (eds.), *Business and Biodiversity*, Satoyama Initiative Thematic Review, https://doi.org/10.1007/978-981-97-7574-3_2

and Development (UNCTAD). Although trade—both domestic and international—is one of the major economic drivers in the world and has the potential to greatly influence the future of the world's biodiversity either negatively or positively, it is an area that has received relatively little attention among those in the biodiversity conservation community to date. BioTrade represents one effort to try to remedy this limited attention by promoting trade in sustainable products—including ensuring that said products are part of a fully sustainable value chain—to bring long-term benefits to communities that produce them in conjunction with its stated goals of contributing to the Convention on Biological Diversity (CBD) and the 2030 Agenda. Through its Principles and Criteria as implemented in communities on the ground, BioTrade can help to promote a SEPLS approach and ensure the long-term economic viability of landscapes and seascapes through sustainable production.

The Regional BioTrade Project in the Mekong region is carried out in Viet Nam, Myanmar, and Lao People's Democratic Republic by Helvetas Swiss Intercooperation (Helvetas) with the goal of "Conservation of biodiversity through sustainable trade of biodiversity products in a manner that integrates local exporters/producers into global value chains and increases income for the rural population of women and men that depend on biodiversity resources for their livelihoods in the Mekong region." It was established in 2016 as a four-year project, and was extended in a second phase to 2024 (Helvetas 2020, 2023). Helvetas has been working with the Duc & Phu Agriculture Joint Venture Company (DPC) to reestablish benzoin gum production in rural villages in Viet Nam (Helvetas 2019).

1.1 The Study Areas

While Siam benzoin gum is produced and sourced in many areas in Laos and northern Viet Nam, this case study will focus on the work of the Regional BioTrade Project in Viet Nam managed by Helvetas Swiss Intercooperation, relying heavily on results from a 2019 study on the effectiveness of this project (Helvetas 2019) as well as materials produced by BioTrade. The purpose of this chapter is to place this work in the context of a SEPLS perspective through literature review, and thus to highlight and analyze its effectiveness as a landscape approach with benefits to nature and human livelihoods in the region. The study looked at collection sites in native national forests in Thanh Hóa and Lào Cai Provinces and plantations in Hòa Bình Province (Fig. 2.1, Table 2.1).

1.2 Siam Benzoin

Siam benzoin gum (Fig. 2.2) is an aromatic balsam obtained from the *Styrax tonkinensis* tree (Fig. 2.3). It is mainly used in the perfume industry, although there is some local use for medicinal purposes. To harvest benzoin gum, trees are tapped by

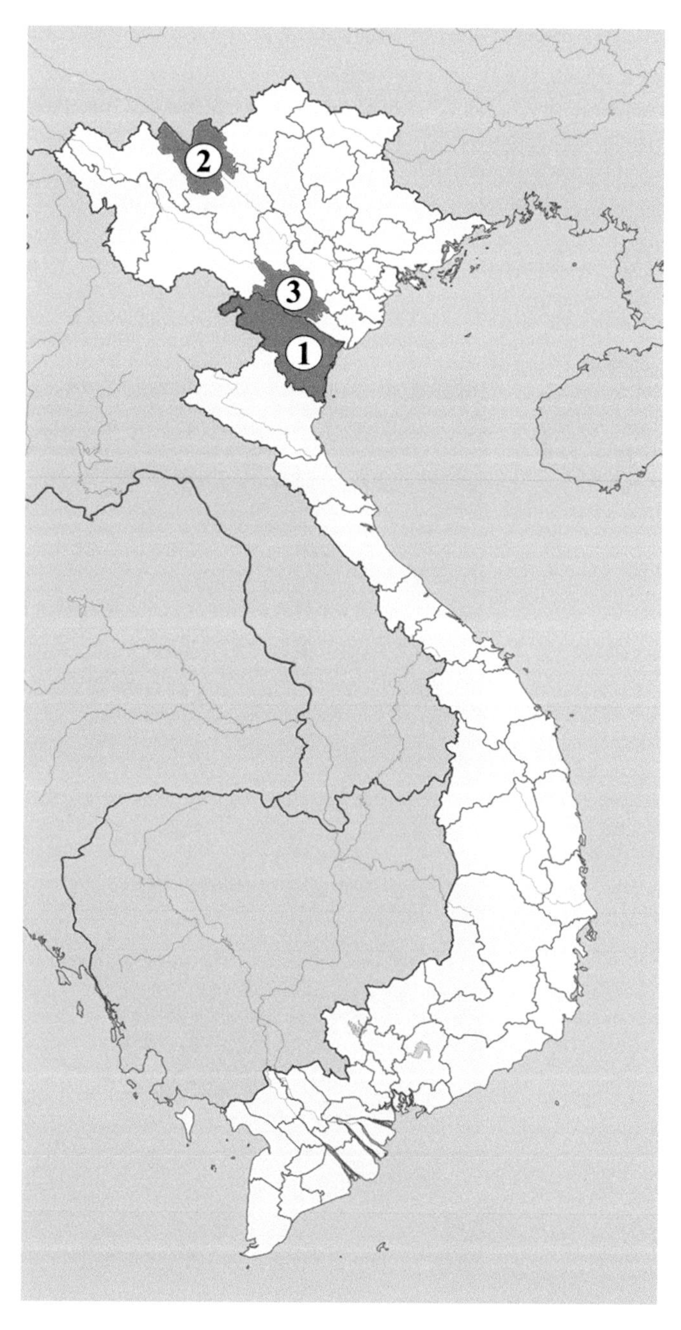

Fig. 2.1 Location of (1) Thanh Hóa, (2) Lào Cai, and (3) Hòa Bình Provinces in Viet Nam. (Source: map data from Wikimedia Commons 2011)

Table 2.1 Basic information of the study area

Country	Viet Nam
Province	Thanh Hóa, Lào Cai, and Hòa Bình Provinces
Dominant ethnicity(ies), if appropriate	Viet (Kinh)
Number of direct beneficiaries (people)	Approx. 1000 households
Geographic coordinates (latitude, longitude)	20°0′N 105°30′E, 22°20′N 104°0′E, 20°20′N 105°15′E

Source: UNCTAD (2016a), adapted by the authors

Fig. 2.2 Siam benzoin gum (Gia Chinh, Helvetas 2023)

Fig. 2.3 Tapping a *Styrax tonkinensis* tree (Gia Chinh, Helvetas 2023)

making cuts in the bark of their trunks once a year, and once the gum is dried it is then collected from the bark. Lao People's Democratic Republic is the largest producer, although it has been produced in Viet Nam as well since the 1990s. Trees are tapped by local residents in the landscape (Fig. 2.3), either wild trees in native national forests or on their own plantations where the trees were mostly initially grown for timber. Trees used for gum harvesting are not cut for timber, meaning that they can remain productive for many years (Helvetas 2019, UEBT 2023b). Benzoin trees were previously grown for logging, but the business became less profitable due to a drop in timber prices, while deforestation became an increasingly serious problem for wildlife and people (UNCTAD 2023c).

The trees also produce small nuts that are thought to be a food source for animals, benefiting the local biodiversity, with Helvetas finding that, "A diverse landscape with *S. tonkinensis* trees forest stands, adjacent to older forests or in very close proximity probably contributes positively to wildlife conservation" (Helvetas 2019). Forest cover in Viet Nam declined drastically from 1943 through 1993, and although reforms have led to an increase in total forest cover in the years since (World Bank 2019), there are concerns about forest quality since much of that increase has been in plantations of non-native cash crops. The *S. tonkinensis* trees are grown in mountainous evergreen mixed closed and broadleaved humid forest, and different sub-types exist at different altitudes. Most production is carried out in middle-range altitudes because lower flatlands are used for agriculture, and the high altitudes are too humid and cold. The production landscapes covered by this study are in close proximity to protected areas, including national parks and nature reserves in Viet Nam and a Laotian national biodiversity conservation area (Helvetas 2019).

1.3 Activities Undertaken

The Regional BioTrade Project in the Mekong region began in 2016, managed by Helvetas and supported by the Swiss State Secretariat for Economic Affairs (SECO). Helvetas has been working with DPC since 2017 to provide technical assistance in collaboration with other stakeholders such as government, private sector, and civil society. This includes activities such as improving management practices, analyzing production methods, and improving better access to international markets. DPC in turn works with over a thousand village households to produce and sell benzoin gum, emphasizing "sustainable use of natural ingredients that contribute to biodiversity conservation and wellness of local communities" (UNCTAD 2023c). For example, workers' wages exceed the national minimum wage, and incentives are paid to producers who conserve and replant the trees. DPC also supports Indigenous and ethnic minority communities, including Mường, Thai, and Dao (Fig. 2.4), through fostering long-term business partnerships with these communities (Woda 2023).

To ensure the economic viability of the benzoin gum sustainable supply chain, the Regional BioTrade Project arranged for DPC to enter into partnership in 2018 with MANE, an international fragrance and flavour sourcing company based in France. This partnership aligned practices along the value chain to the BioTrade Principles and Criteria (P&C) for sustainable and biodiversity-friendly sourcing of

Fig. 2.4 Dao women working on collection process of benzoin (Gia Chinh, Helvetas 2023)

benzoin gum (UNCTAD 2023c). The logic of the value chain is essentially that buyers are motivated by sustainable management of the raw resources and fair working conditions for producers in order to realize sustainable sourcing, while producers prioritize fair payment, occupational safety, quality, and traceability. Ideally, this results in a situation in which everyone, including nature, wins (Helvetas 2019).

The project's work is potentially made somewhat more complex by policies of the Forest Protection Board (FPB)—the state forest administration that is responsible for setting management regulations and issuing permissions for forest resource use (Helvetas 2019)—which has not effectively regulated non-timber forest products (NTFPs) but in recent years enforced a total logging ban on native species. As a result, some species used for medicinal and other purposes may have been overexploited to the point of extinction, while local communities are unable to effectively manage the population of trees including those used for Siam benzoin. Research by Helvetas has shown that local communities have good knowledge for management of these trees, which have a lifespan of around 60 years and need to be periodically cut down and replanted. A blanket ban on cutting these trees would likely result in the trees being replaced by other species through forest succession, causing the local communities to lose this source of income. Notably, these rules do not apply to plantations, meaning benzoin would only be produced where wild forests have been cut down and planted with the cash crops. For this reason, ongoing work by BioTrade includes identifying partners who can advocate for key actors to promote sustainable approaches to forest management in the affected areas (Helvetas 2019). More information about how BioTrade works with partners and its working principles is presented in the following section.

2 BioTrade

This section presents an overview and details of BioTrade's concepts and work with various projects around the world, with a focus on how these relate to the Siam benzoin gum project in Viet Nam. BioTrade was established on the idea that conservation of biodiversity and the sustainable use and trade of biodiversity-based products and services can provide countries valuable opportunities for economic development and improvement of livelihoods. UNCTAD launched the BioTrade Initiative in 1996 to support the three objectives of the CBD, namely, the conservation of biological diversity, the sustainable use of its components, and the fair and equitable sharing of the benefits arising out of the utilization of biodiversity (CBD 1992). Trade is in a unique position to bring together all three objectives in that human production activities are one of the major drivers of biodiversity loss. As such, making production and consumption sustainable throughout the value chain could have a major positive effect, while sustainable systems of trade can bring long-lasting community benefits to local communities contributing to equitable benefit sharing. Since its founding, the initiative has developed a unique portfolio of global, regional, and country programmes as well as a network of partners and practitioners working in over 80 countries (UNCTAD 2022).

BioTrade has been implemented for over 25 years. Starting in the late 1990s and early 2000s, the initiative defined the concept of BioTrade and how it is implemented on the ground to ensure the sustainability of its interventions. In this work, it pursued innovative collaborative arrangements to develop sustainable trade in biodiversity-based products and services and supported developing countries in accessing new markets. At the same time, UNCTAD has also focused on creating a policy environment that would promote trade and investment based on sustainable use of biodiversity. These efforts helped to create jobs, income, export diversification, and rural development for enterprises and organizations engaged in the sector, and enhanced the livelihoods of rural and local communities in developing countries with economic, environmental, and social benefits. As the initiative evolved, UNCTAD has also worked to mainstream BioTrade and its experiences and lessons learned into multilateral, regional, and national processes. They have also been strengthening the policy and regulatory environment, and increasing their involvement in international policy processes such as compiling and analyzing non-tariff measures (NTMs) in key import and export markets, assessing the potential implementation of traceability systems for targeted CITES species, and mapping and providing recommendations for implementation of the Nagoya Protocol (UNCTAD 2016b).

2.1 BioTrade Principles and Criteria

Fundamentally, BioTrade aims to ensure that the commercialization and trade in products and services sourced from biodiversity are done in a way that respects people and nature. The UNCTAD BioTrade Principles and Criteria (P&C) were developed jointly with partners and published in 2007, followed by an update in 2020. These P&C are a set of guidelines for businesses, governments, and civil society wishing to support the objectives of the SDGs and other multilateral environmental agreements such as CBD and CITES. The P&C have been implemented

Table 2.2 The BioTrade Principles and Criteria, as of 2020

Principles	Criteria
1. Conservation of biodiversity	1.1 Activities contribute to maintaining, restoring, or enhancing biodiversity, including ecosystems, ecological processes, natural habitats, and species, particularly threatened or endangered species
	1.2 Genetic variability of flora, fauna, and microorganisms (for use and conservation) is maintained, restored, or promoted.
	1.3 Activities are aligned with national, regional, and/or local plans for sustainable management, conservation, and restoration of biodiversity, in coordination with the relevant authorities and actors involved
2. Sustainable use of biodiversity	2.1 The use of biodiversity is sustainable, based on adaptive management practices that advance the long-term viability of the biological resources used, and supported by training of workers and producers on good collection, harvesting, cultivation, breeding, or sustainable tourism practices
	2.2 Measures are taken to prevent or mitigate negative environmental impacts of the activities, including in relation to flora and fauna; soil, air, and water quality; the global climate; use of agro-chemicals; pollution and waste disposal; and energy consumption.
	2.3 Activities contribute to measures that strengthen resilience and the adaptive capacity of species and ecosystems to climate-related hazards and natural disasters
3. Fair and equitable sharing of benefits derived from the use of biodiversity	3.1 Activities are agreed upon and undertaken based on transparency, dialogue, and long-term partnerships between all organizations involved in the supply chain.
	3.2 Prices take into account the costs of value chain activities (e.g. production, investment, research and development, marketing, and commercialization) according to these Principles and Criteria and allow for a profit margin.
	3.3 Activities contribute to sustainable local development, as defined by producers and their local communities
	3.4 Activities comply with applicable legal requirements and/or relevant contractual arrangements on access to biodiversity, including biological and genetic resources, their derivatives and associated traditional knowledge, and on the fair and equitable sharing of benefits derived from their utilization
	3.5 In cases where there are no applicable legal requirements, utilization of genetic resources and associated traditional knowledge takes place with prior informed consent and mutually agreed terms

(continued)

Table 2.2 (continued)

Principles	Criteria
4. Socio-economic sustainability	4.1 The organization demonstrates the integration of these principles and criteria in its business and supply chain management
	4.2 The organization has a quality management system in line with its market requirements.
	4.3 A system is in place to allow for supply chain traceability up to the country of origin and/or the place of collection, harvesting, and/or cultivation
5. Compliance with national and international regulations	5.1 The organization complies with applicable legal and administrative requirements at local, national, and regional levels. If measures required by local, national, or regional legislation are less strict than those required by these principles and criteria, the organization meets the stricter requirements
	5.2 Activities respect the principles and obligations of relevant international agreements and instruments, such as the CBD, Nagoya Protocol, International Treaty on Plant Genetic Resources for Food and Agriculture (ITPGRFA), Convention on International Trade in Endangered Species of Wild Fauna and Flora (CITES), Bonn Convention on Migratory Species (CMS), International Labour Organization (ILO) Conventions, United Nations Declaration on the Rights of Indigenous Peoples, and United Nations Declaration on the Rights of Peasants and Other People Working in Rural Areas
	5.3 When dealing with marine and coastal biodiversity, activities respect the principles and obligations established under the United Nations Convention on Law of the Sea (UNCLOS), United Nations Fish Stocks Agreement (UNFSA), and any subsequent instrument on biodiversity in areas beyond national jurisdiction, as well as relevant conventions and instruments adopted under the United Nations Conference on Trade and Development (UNCTAD), Food and Agricultural Organization of the United Nations (FAO), UN Environment, International Maritime Organization (IMO), and International Labour Organization (ILO)
	5.4 The organization gathers and maintains information and records required to ensure the legality of access to and use of biodiversity, such as the country of origin, geographical location of capture or introduction from the sea, existence of applicable laws or regulations, and relevant permits and certificates
6. Respect for the rights of actors involved in BioTrade	6.1 The organization respects fundamental human rights, in keeping with the United Nations Guiding Principles on Business and Human Rights and relevant ILO Conventions
	6.2 The organization respects worker rights, provides adequate working conditions, and prevents any negative impacts on the health and safety of workers, in accordance with national legislation
	6.3 The organization respects the rights of Indigenous peoples and local communities, women, children, and other vulnerable groups involved in BioTrade activities, in accordance with national legislation and the United Nations Declaration on the Rights of Indigenous Peoples

(continued)

Table 2.2 (continued)

Principles	Criteria
7. Clarity on the right to use and access to natural resources	7.1 The organization uses natural resources in compliance with all relevant laws and regulations and prevents any negative impacts on the health, safety, and well-being of surrounding populations
	7.2 In cases where required by international, national, local, or customary law, as well as Criteria 3.5, the organization accesses natural resources and associated traditional knowledge with prior informed consent of, and subject to mutually agreed terms with, the party that provides them
	7.3 The organization respects the rights of Indigenous peoples and local communities over land, natural resources, and associated traditional knowledge in accordance with national legislation and the United Nations Declaration on the Rights of Indigenous Peoples
	7.4 The organization does not threaten the food diversity or food security of producers and their local communities

Adapted from: UNCTAD (2021)

since 2007 by governmental organizations, NGOs, and the private sector in nearly 100 countries. The BioTrade P&C are shown in Table 2.2 below.

The Siam benzoin gum project in Viet Nam supports these P&C (UNCTAD 2023c) and contributes to conservation of biodiversity (Principle 1) based on the benefits mentioned above, namely by preventing the deforestation of native species and providing a food source for animal species. Fair and equitable sharing of benefits (Principle 3) and respect for the actors involved (Principle 6) are at the heart of the project, as it is fundamentally a biodiversity-based livelihoods project with the goal of improving local well-being through sustainable trade (Helvetas 2023).

2.2 *Relation to Other BioTrade Projects*

For more context on BioTrade's work in promoting Siam benzoin gum in Viet Nam, below are some other examples of how UNCTAD and its partners implement the BioTrade P&C in promoting products around the world, and the landscape benefits that arise. While these projects are not directly related to Siam benzoin, they underscore how biodiversity conservation and livelihood benefits can be derived through the BioTrade P&C.

Similar benefits related to NTFPs, but with a stronger emphasis on Principles 2 (sustainable use of biodiversity) and 4 (socioeconomic sustainability), can be seen in Mozambique, where the baobab tree (*Adansonia digitata* L.) typically grows in arid areas and holds cultural and traditional significance. Almost all parts of the tree can be used including its bark, leaves, and seed, and the pulp of its fruit has been gaining international attention as a superfood. Baobab collection is a source of income for local communities that is threatened by a decrease in the trees caused by deforestation and climate change. With demand increasing globally, consistent sourcing has been a challenge as international buyers demand more stringent standards. Local partners formed the shareholding company Baobab Products Mozambique (BPM) to collect, process, and market fruit powder, and it has adopted the BioTrade P&C for sustainable harvesting methods. BPM employs over 2000 women workers, enabling these women to invest in household assets, small businesses, and access to services. Every sack of powder can be traced back to its sourcing and production, and collectors are trained in quality control so that products meet international market standards. BPM has invested in long-term relationships with distributors who share the BioTrade vision, and these have facilitated a more inclusive business and made the company more resilient in the international market (UNCTAD 2023a).

BioTrade's experience promoting plant products for the sustainable use of biodiversity (Principle 2) and fair and equitable sharing of benefits (Principle 3) informs its work in the three rural communities of La Carbonera, La Joya, and Charape-La Joya in Mexico. Knowledge of local aromatic and medicinal plants has been passed down for generations and has great economic potential, but the area has suffered from lack of economic opportunities due to lack of capacity, the arid climate, and soil degradation. Provital is a Spanish cosmetics company and a Union for Ethical BioTrade (UEBT) member. They have been working with Mujeres y Ambiente (M&A), a women's community-based initiative that offers training to women on combining methods of sustainable agriculture with traditional knowledge. Together, they have worked to develop integrated and sustainable watershed landscapes in the communities and started training the communities on how to manage and conserve the watershed. In 2016, an Access and Benefit Sharing (ABS) Agreement was signed which confirms that the community will receive monetary and non-monetary benefits. Direct and indirect benefits support 60 families in the region, and the sustainable water management measures introduced benefit biodiversity. M&A is now working to upscale the project by offering training to groups of women producers in other communities. In addition to sustainable use and benefit sharing, this project is a good example of the BioTrade Principle 5 on compliance with national and international regulations (UNCTAD 2023b).

2.3 *The Union for Ethical BioTrade*

The Union for Ethical BioTrade (UEBT) is a not-for-profit association founded in 2007 that promotes business engagement in the ethical sourcing of biodiversity with the stated goal of "sourcing with respect" through its Ethical BioTrade Standard. This standard is based on the BioTrade Principles and Criteria. UEBT works by guiding companies and their suppliers in how to harvest, collect, or grow ingredients from biodiversity in a way that respects people and biodiversity (UEBT 2023a). It also raises awareness on ethical sourcing of biodiversity through its membership programme, has a knowledge-sharing platform, and organizes international conferences and workshops. It verifies ethical sourcing through its two certification programmes for ingredients from biodiversity and for ethical sourcing systems (UEBT 2020b). DPC has also been a UEBT member since 2020.

UEBT sets practices for how companies and their suppliers produce, develop, process, and purchase ingredients from biodiversity for use in food, cosmetics, and natural pharmaceutical products, based on its Ethical BioTrade Standard. This standard serves as a practical guide to adherence to the BioTrade P&C for businesses and other organizations and establishes good practices along the supply chain. The standard extends from cultivation and wild collection to the UEBT companies, farms, processors, manufacturers, and brands. It aims to improve conservation of biodiversity, sustainable use, fair prices, income, access and benefit sharing, rights, local development, and food security (UEBT 2020a).

UEBT also works with companies to develop Biodiversity Action Plans (BAPs), which serve as comprehensive frameworks to guide and support member companies in sourcing biodiversity-based ingredients and products in an ethical and sustainable manner. UEBT, in collaboration with Helvetas, has been working with DPC to develop their own BAP, which resulted in getting the benzoin value chain certified in 2023.

In 2020, UEBT published a set of lessons learnt from its first 10 years of work, which showed the extent to which biodiversity is both an ecological and an economic imperative for business that relies on biodiversity for its long-term viability. It also found that consumers expect biodiversity-friendly practices, that biodiversity requires on-the-ground engagement, and that people and biodiversity are inherently linked. In this regard, standards are effective tools for biodiversity action, and biodiversity must be an integral part of company strategies, operations, and supply chains. At the same time, it is necessary to recognize access and benefit-sharing rights over biodiversity and to tailor biodiversity actions to the local context. With this background, UEBT is looking to further work in partnership with its member organizations and beyond, and also to work towards biodiversity regeneration in addition to conservation (UEBT 2020b).

3 Results

3.1 Biodiversity and Community Impacts

Helvetas research indicates that the biodiversity benefits of the *Styrax tonkinensis*, while difficult to measure directly, derive from the inherent conservation potential of the species as well as the nature of the production system surrounding it. Further factors are the protection needs of the ecosystem and the actions of those in the supply chain, including regulators. In this sense, the tree is shown to have inherent biodiversity benefits from its place in the forest succession and its use as a food source for local fauna, as described above. Since the benzoin gum represents an obvious income source for the local communities that produce it based on conservation and maintenance of the trees, the production system also can be said to work effectively towards biodiversity conservation. It is important to note that in this particular ecosystem, natural forest succession would mean that these trees would be replaced by other species in time, so that some degree of forest management is needed to maintain the trees and the resource. In this way, management has real biodiversity benefits in the ecosystem. Tellingly, among the sites covered by the Helvetas study, Thanh Hóa Province has the most protected areas, and benzoin gum production takes place in the closest proximity to these areas. It also is the most productive area, indicating that conservation benefits and production effectiveness can be mutually beneficial when the production system is aligned with buffer zones or green corridor management (Helvetas 2019).

Due to the trees' proximity to protected natural areas, they are home to many animals, including endangered species such as the northern white-cheeked gibbon (*Nomascus leucogenys*), wild dog (*Cuon alpinus*), and the Chinese pangolin (*Manis pentadactyla*). Due to high agricultural activity in the lower valley areas, benzoin trees have a strong potential to create buffer zones or be used for green corridor management, which can contribute to biodiversity conservation (Woda 2023).

Nevertheless, there is no formal collaboration among DPC, producers, the Forest Protection Board, and the local authorities. Better communication channels are needed to leverage the potential for benzoin gum production and management of benzoin trees to have a positive impact on biodiversity conservation and support the livelihoods of local communities (Woda 2023).

In terms of economic benefits to the community, local people have been strongly motivated to maintain the trees through benzoin gum production (Fig. 2.5). When the trees were grown primarily for logging, a tree would generate income of around USD 8.50 when it was cut down after reaching maturity, while a mature tree used for benzoin gum can yield USD 11.00 each season. A related benefit is income diversification in communities where the main source of income is agriculture, helping to offset fluctuations in agricultural commodity prices. DPC has also been working to ensure that workers are treated well and has improved occupational safety measures in place (UNCTAD 2023c).

Fig. 2.5 Local benzoin collector (Gia Chinh, Helvetas 2023)

Elsewhere along the value chain, the BioTrade-facilitated partnership provides the buyer MANE with a more reliable supply of high-quality benzoin gum that also meets its sustainability standards. Meanwhile, the FPB's goal of protecting the nation's forests could potentially benefit from this kind of sustainable production activity, but awareness of its effectiveness remains low, and current policies including the cutting ban mentioned above, may inadvertently lead to negative impacts (Helvetas 2019).

3.2 Ongoing and Future Work

While the second phase of the Regional BioTrade Project is still ongoing until 2024, Helvetas has identified some priorities for the future, including creation of a clear action plan for conservation of the trees and stronger coordination with local governments. There is a perceived gap between the strict regulations enforced on timber activities and those for NFTPs like benzoin gum, and a clear plan for proactive steps is now being developed (UNCTAD 2023c).

The need for more coordination with the FPB is also highlighted by the Helvetas study, which recommends establishing a platform for closer communication and investigation of knowledge on the biodiversity benefits of the trees and best practices for management of the forest landscape. The next step may also be to look into the possibilities for restoration—not only conservation—of the landscape, combining BioTrade sustainable production practices with others such as those promoted by IUCN. The ultimate goal of all of these improvements is to work with the local communities to ensure long-term viability of the forest and the sustainable supply chain of benzoin gum (Helvetas 2019).

4 Conclusions

With the ongoing global trends in biodiversity loss poised to seriously affect businesses that rely on biological and genetic resources, those working with pharmaceutical and related resources are already being recommended to plan for changes in biodiversity and more stringent benefit-sharing requirements (TEEB 2012). The BioTrade approach illustrated in Sect. 2 of this chapter is self-described as an attempt to contribute to the three objectives of the Convention on Biological Diversity through promotion of sustainable production, sourcing, and trade of biodiversity-based products and services. Alternatively, the Regional BioTrade Project in the Mekong region puts it as, "conservation of biodiversity through sustainable trade of natural ingredients in a manner that increases the competitiveness of local exporters/producers and the livelihood benefits (income and jobs) of rural women and men" (Helvetas 2019). In short, the approach is closely aligned with the SEPLS-based landscape approach promoted under the Satoyama Initiative.

This chapter presents the BioTrade project for the production, processing, and marketing of Siam benzoin gum by DPC in three provinces of northern Viet Nam, supported under the Regional BioTrade Project led by Helvetas. Under this project, local communities' management of the trees that produce the benzoin gum has a positive impact on biodiversity, since the trees themselves are a food source for local fauna and because periodic cutting and replanting of the trees for economic purposes prevents them from being replaced by other species in the process of forest succession. Moreover, the local producers are financially motivated to continue managing the trees over the long term because sales of the benzoin gum provide additional income as well as income diversity, contributing to resilience.

Up to this time, the focus in the benzoin gum market appears to have primarily been at the species level. However, recent trends in the conservation world have made it more apparent that not only the individual tree species but the whole forest landscape with its various ecosystems ought to be considered in this and related projects, as the landscape produces not only biodiversity but also livelihood benefits. There are legal policies and frameworks currently in place in Viet Nam to enforce forest management, such as the Forestry Law, which regulates the management, protection, strategy, and development of the use of forests in Viet Nam (Viet Nam 2017), while protected areas and national parks are guarded by forest rangers. Nevertheless, it is evident that projects such as this one are crucial platforms for local communities, regulatory agencies, conservation NGOs, and others to cooperate and understand the needs of the ecosystem better, and also to promote forest landscape restoration efforts in line with international standards.

It is only in the past few years that UNCTAD's BioTrade Initiative has come into contact with the Satoyama Initiative, and a SEPLS perspective is not yet a major part of its operations. Nevertheless, the similarity in approach is undeniable, and there is a great deal of potential for greater effectiveness in terms of conservation, equity, and other priorities through incorporation of landscape approaches as appropriate. Conversely, trade may not have received as much attention as some other issues under the Satoyama Initiative, and UNCTAD BioTrade's experience with sustainable trade could have a lot to share with those who have not considered it in the past. Trade is one of the indirect drivers of biodiversity loss (IPBES 2019) and loss of resilience in the world, and this approach shows great potential to reduce its negative impacts and even turn it into a net positive for nature.

References

Convention on Biological Diversity (CBD) (1992) Convention on Biological Diversity

Helvetas (2019) Case Studies: Biodiversity Impact Learning from the cases of Siam benzoin, organic chili, Indian prickly ash and jujube in Vietnam, Laos and Myanmar. Helvetas, Geneva

Helvetas (2020) Regional BioTrade Project in South East Asia: Phase II. Helvetas, Geneva

Helvetas (2023) Vietnam, Laos, Myanmar: Ethical Trade in Botanicals, viewed 20 February 2023. Retrieved from https://www.helvetas.org/en/switzerland/what-we-do/how-we-work/our-projects/asia/vietnam/vietnam-laos-myanmar-regional-market

IPBES (2019) Summary for policymakers of the global assessment report on biodiversity and ecosystem services of the Intergovernmental Science-Policy Platform on Biodiversity and Ecosystem Services. IPBES secretariat, Bonn. https://doi.org/10.5281/zenodo.3553579
TEEB (2012) The Economics of Ecosystems and Biodiversity in Business and Enterprise, Ed. Joshua Bishop. Earthscan, London and New York
UEBT (2020a) Ethical BioTrade Standard. UEBT, Amsterdam
UEBT (2020b) The Big Shift: Business For Biodiversity – Lessons Learned from over 10 years of the Union for Ethical BioTrade (UEBT). UEBT, Amsterdam
UEBT (2023a) Who we are, viewed 1 March 2023. Retrieved from https://uebt.org/about-uebt
UEBT (2023b) Benzoin: Sweet tears from the snowbell tree, viewed 20 October 2023. Retrieved from https://uebt.org/ingredient-stories/benzoin
UNCTAD (2016a) 20 Years of BioTrade: Connecting People, the Planet and Markets. UNCTAD, New York and Geneva
UNCTAD (2016b) The Interface Between Access and Benefit-sharing Rules and BioTrade in Viet Nam. UNCTAD, New York and Geneva
UNCTAD (2021) Implications of the African Continental Free Trade Area for Trade and Biodiversity: Policy and Regulatory Recommendations. UNCTAD, New York and Geneva
UNCTAD (2022) BioTrade, New York and Geneva, viewed 18 February 2023. Retrieved from https://unctad.org/topic/trade-and-environment/biotrade
UNCTAD (2023a) Building an Inclusive Baobab Value Chain in Mozambique. UNCTAD, New York and Geneva (forthcoming)
UNCTAD (2023b) Harvesting and Protecting Traditional Plants in Mexico. UNCTAD, New York and Geneva (forthcoming)
UNCTAD (2023c) Siam Benzoin Gum in Viet Nam. UNCTAD, New York and Geneva
Viet Nam (2017) Forestry Law, Law No. 16/2017/QH14. https://faolex.fao.org/docs/pdf/vie206322.pdf
Wikimedia Commons (2011) Location of Thanh Hóa within Vietnam, viewed 20 February 2023. Retrieved from https://en.wikipedia.org/wiki/Thanh_H%C3%B3a_province#/media/File:Thanh_Hoa_in_Vietnam.svg
Woda C (2023) Biotrade Case Studies: Siam Benzoin Gum in Vietnam, Helvetas, Geneva (forthcoming)
World Bank (2019) Forest Country Note – Vietnam. World Bank, Washington, DC

William Dunbar Project Manager at Conservation International in Tokyo, Japan, focusing on international sustainable landscape and seascape policies, particularly relating to landscapes as “other effective area-based conservation measures” (OECMs).

Lorena Jaramillo Economic Affairs Officer and the project manager of the Global BioTrade Programme at UNCTAD Switzerland, working in areas of trade and biodiversity, sustainable trade, and business development.

Lika Sasaki Programme Management Officer at UNCTAD Switzerland for the Global BioTrade Programme. BioTrade promotes the sustainable trade of biodiversity-based products and services in a way that can benefit nature and sustain livelihoods.

BY NC SA

Chapter 3
Chestnut Production-Related Businesses in the Courel Mountains of Galicia, NW Spain: An Opportunity for Biodiversity Conservation, Ecosystem Restoration, and Rural Development

Emilio R. Díaz-Varela, Pedro Álvarez-Álvarez, José C. Pérez-Girón, Ramón A. Díaz-Varela, María A. Ferreira Golpe, and Ana I. García-Arias

Abstract Chestnut orchards (*soutos* in the Galician language) constitute the main feature of socio-ecological production landscapes (SEPLs) characteristic to the Northwest Iberian Peninsula. These landscapes display high levels of biodiversity that are shown in the genetic, interspecific, and ecosystem domains. They also produce a variety of ecosystem services, including cultural, regulating, and provisioning ones. Nevertheless, the interrelation between ecological functions, ecosystem services, and businesses in the present day is threatened by the abandonment of traditional management, giving way to a range of transformations that affect both the mountain landscapes and their socio-economic fabric.

The objective of this study is to explore the relationships between socio-economic activities and the capacity of chestnut SEPL to preserve high biodiversity levels in the municipality of Folgoso do Courel (Galicia, NW Spain), an area with a strong tradition in chestnut production systems. To do so, we first analyse the economic structure of the area using official statistics and identifying specific businesses related to the local SEPL. We also use a geographical information system to analyse land cover maps to locate and characterize chestnut production areas. We complete our analysis by interviewing business owners to elicit important elements in the

E. R. Díaz-Varela (✉) · R. A. Díaz-Varela · M. A. Ferreira Golpe · A. I. García-Arias
Higher Polytechnic Engineering School, University of Santiago de Compostela, Lugo, Galicia, Spain
e-mail: emilio.diaz@usc.es

P. Álvarez-Álvarez
Polytechnic School of Mieres, University of Oviedo, Mieres, Asturias, Spain

J. C. Pérez-Girón
Polytechnic School of Mieres, University of Oviedo, Mieres, Asturias, Spain

Department of Botany, Faculty of Sciences, University of Granada, Granada, Spain

Interuniversity Institute for Earth System Research in Andalusia (IISTA), University of Granada, Granada, Spain

M. Nishi et al. (eds.), *Business and Biodiversity*, Satoyama Initiative Thematic Review, https://doi.org/10.1007/978-981-97-7574-3_3

business-SEPL relationship. Our results show the strong dependence of chestnut-related businesses on the local SEPL, and how the multifunctional aspects of chestnut production are important assets in the businesses' visions. Also, we identify risks and impacts affecting the socio-ecological production landscape. We conclude that supporting SEPL-related businesses could benefit biodiversity conservation and sustainability in the territorial system.

Keywords Traditional chestnut orchard management · Multifunctional agroecosystem management · Mountain areas · Ecosystem services · Conservation of traditional knowledge · Socio-ecological production landscapes

1 Introduction

1.1 Agroecosystems, Related Business, and Sustainable Development

The current global effort to achieve sustainability demands the participation of all sections of society in a variety of initiatives aimed at the integration of biodiversity conservation, climate change mitigation, and adaptation, and the flow of ecosystem services favouring inclusive development. These global imperatives demand both transversal and vertical integration for management of ecosystems and the products and services obtained from them. The Kunming-Montreal Global Biodiversity Framework (GBF) adopted at the 15th meeting of the Conference of the Parties to the Convention on Biological Diversity (CBD) in Montreal, Canada, in December 2022 includes maintenance and enhancement of the resilience, integrity, and connectivity of natural and sustainably managed ecosystems among its goals and targets. In rural areas, agroecosystems, forests, and agroforestry systems are essential providers of a variety of nature's contributions to people, securing livelihoods apart from conserving biodiversity and cultural heritage, mitigating climate change, and rendering socio-ecological resilience (Garbach et al. 2014; Barral et al. 2015).

Traditional sweet chestnut (*Castanea sativa* Mill.) orchards, called *soutos*, are a unique distinguishing feature of socio-ecological production landscapes (SEPLs) in Galicia and other neighbouring regions of the NW Iberian Peninsula. Their historical development in the Courel Mountains can be traced back centuries (Pereira-Lorenzo et al. 2019; Fernández-López et al. 2021). At present, their management is based on a combination of traditional and modern practices. They display high levels of biodiversity that are shown in the genetic (their long cultural history induced profuse intraspecific variety), interspecific (being a suitable habitat for many species, including umbrella species like brown bear), and ecosystem (together with other ecosystems, they contribute to the ecological integrity of landscapes) domains. Such biodiversity constitutes the functional basis for producing ecosystem services, including cultural, regulating, and provisioning services. The latter are relevant to businesses at both the regional and local level: the production of chestnut fruits (up to 3000 kg per hectare annually), mushrooms (up to 200 kg per hectare annually), honey, timber, and livestock, as well as landscape-related tourism. These businesses are highly important, both directly and indirectly, to the integrity of the SEPLs.

Galicia accounts for 66% of Spain's chestnut production area and 90% of production (MAPA 2021). Galicia produced 81,084 tonnes of chestnut with a market value of 20 million EUR in 2020 (Consellería de Medio Rural 2022b). However, only 5.2% of the chestnut area in Galicia is under a regional-level sectoral protection and quality certification institution called Protected Geographical Indication (PGI), which brands the product as *Castaña de Galicia* and is the only quality label for chestnuts in Spain (Consellería de Medio Rural 2021). Certified production was 24.9 tonnes in 2020, 67.7% fresh and 32.3% frozen, and 28.6 tonnes in 2021. This represents a significant decrease in production compared to 2016, when production under PGI amounted to 240 tonnes. Reasons behind this decrease are the economic effects of the COVID-19 pandemic, together with the impact of certain pests and diseases, such as the invasive *Dryocosmus kuriphilus* wasp and the *Gnomoniopsis castanea* fungi (Lombardero et al. 2022; Consellería de Medio Rural 2022a; Aguín et al. 2023). The estimated value of this certified production for 2021 amounted to 233,000 EUR. Regarding the organization of the economic sector, PGI currently accounts for 169 producers, 10 sellers, and 6 processing industries. Most products are locally consumed. Only 3% of total product was exported in 2021, and 51% of shelled chestnuts were utilized in the food industry. Despite its significant production, Galicia imported 1306.7 tonnes of chestnut in 2021 with a value of 2.6 million EUR. Imports have been increasing since 2016 while exports are decreasing, following the trend of production (MAPA 2022).

Researchers have paid attention to *soutos* largely in terms of their production function, ignoring the complexity of eco-agri-food systems (Zhang et al. 2018). Despite the *souto*'s importance in the various dimensions of sustainability, comprehensive knowledge on their multifunctionality and direct and indirect interactions with other ecosystems in SEPLs is lacking. These interactions manifest as (a) the *risks* from ecological degradation to economic activity (e.g. decline in production from forest fires and loss of regulating services keeping pests/pathogens in check and favouring pollinators and beneficial soil organisms); (b) the ecological *impacts* of economic activities (e.g. air, soil, and/or water pollution, overexploitation of natural resources, unsustainable tourism), with obvious effects not only on ecosystem functioning and ecological integrity, but also on business and governance; and (c) new business *opportunities* including those related to ecosystem restoration, enhanced efficiency in the use of natural resources and ecosystem services, or benefits associated with the acknowledgement of good environmental practices or production of ecosystem goods and services through payment schemes (WBCSD 2011; Hanson et al. 2013).

Nevertheless, the interrelation between ecological functions, ecosystem services, and business is nowadays threatened by the abandonment of traditional management, which has given way to a range of transformations affecting both the mountain landscapes and their socio-economic fabric. These include intensive afforestation practices, industrial slate quarrying, and wind farming. They have induced irreversible changes not only in land use, but also in the general productive arrangement of the territory from a regenerative to an extractive one, with clear consequences for the sustainability of the whole system. The abandonment of traditional management might also lead to a shift towards a different habitat structure and composition due to natural ecological succession or even, in the case of intensified production systems, to a simplification of the habitat hosting lower biodiversity values.

1.2 Objectives

The objective of the present research was to explore the relationships between socio-economic activities and the capacity of a chestnut-dominated SEPL to conserve biodiversity in the Municipality of Folgoso do Courel (Galicia, NW Spain). This study is a component of a wider research programme on socio-ecological characterization of *soutos*-related rural systems in the NW Iberian Peninsula and complements the "CASTEXEN: Location and differentiation of chestnut (*Castanea sativa* Mill.) grafted with traditional cultivars using spatial-temporal analysis of aerial photographs and SSR genotyping" research project, financed by the Campus Terra of the University of Santiago de Compostela (Spain). As such, it is exploratory work aiming for a better understanding of the socio-economic subsystem of the chestnut SEPL.

2 Material and Methods

2.1 Study Area

Folgoso do Courel is a rural municipality located in the Courel Mountains (Galicia, Spain; see Fig. 3.1 for location, and Fig. 3.2 for basic geographical features). It is characterized by diverse geological substrata shaped in convoluted geomorphology features, recently acknowledged through the designation of the Courel Mountains UNESCO Global Geopark. The area experiences a mix of temperate and Mediterranean climate attributes and is highly rich in biodiversity at the genetic, species, and ecosystem levels. The endangered brown bear (*Ursus arctos*) is an umbrella species. Almost the entire municipality is protected under the European Union "Natura 2000" network, with the exception of a slate quarry area in the south. The population is ageing and decreasing in numbers. Economic activities in the primary sector include livestock and forestry production, which depend largely on local ecosystems (Table 3.1).

The traditional SEPL is a mosaic of traditional agricultural patches, shrublands, and chestnut orchards. Fire risks have increased as a result of abandonment of traditional practices (e.g. tree pruning, forest clearing, and understorey livestock

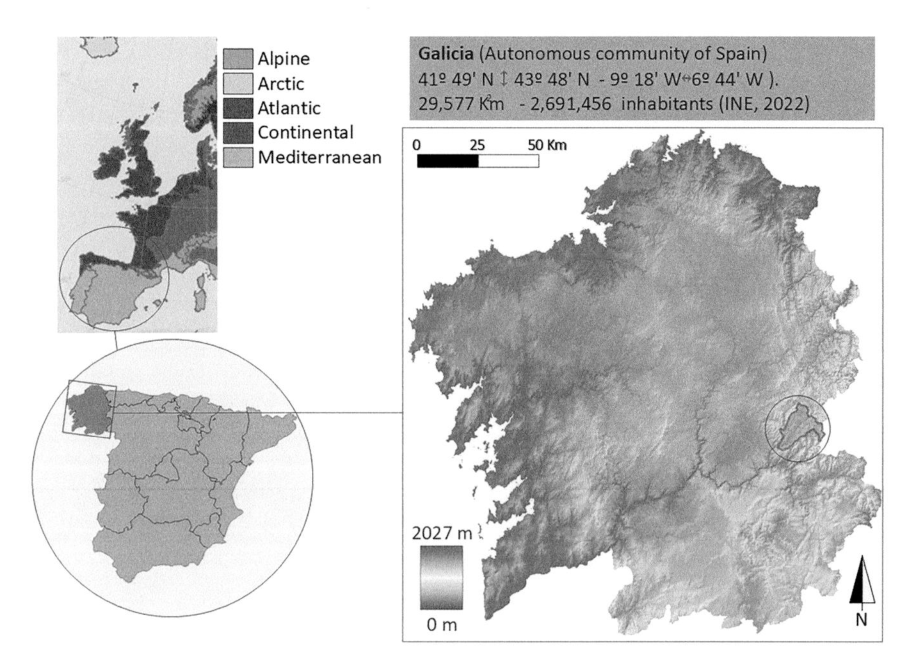

Fig. 3.1 Study area in Galicia, Spain. (Source: Prepared by authors; Digital terrain model and other GIS data from Centro Nacional de Información Geográfica (2019); Biogeographical map from European Environment Agency (2017); Demographic data from INE (2022))

grazing) and introduction of altogether new ones like intensive afforestation. Pressures from industrial activities include the expansion of slate quarries and planning of future wind farms.

2.2 *Methodological Approach*

A series of steps were taken to explore the connections in Folgoso do Courel between socio-economic activities and the capacity of the chestnut SEPL to conserve biodiversity and provide ecosystem services.

1. First, we analyzed the economic activities in the area and businesses associated with *soutos*. General features were assessed through official statistics available from the Galician Institute for Statistics (IGE 2023) and the Ministry for Rural Environment (Consellería de Medio Rural 2022b). Specific business orientations were identified through interviews with members of the Regulation Council of the Protected Geographical Indication "Galician Chestnut" (Indicación Xeográfica Protexida Castaña de Galicia; Consellería de Medio Rural 2021). The Protected Geographical Indication (PGI) scheme based on European legislation (Regulation (EU) No 1151/2012 and Commission Regulation (EU) No 409/2010) secures fair returns for the producers and, at the same time, ensures

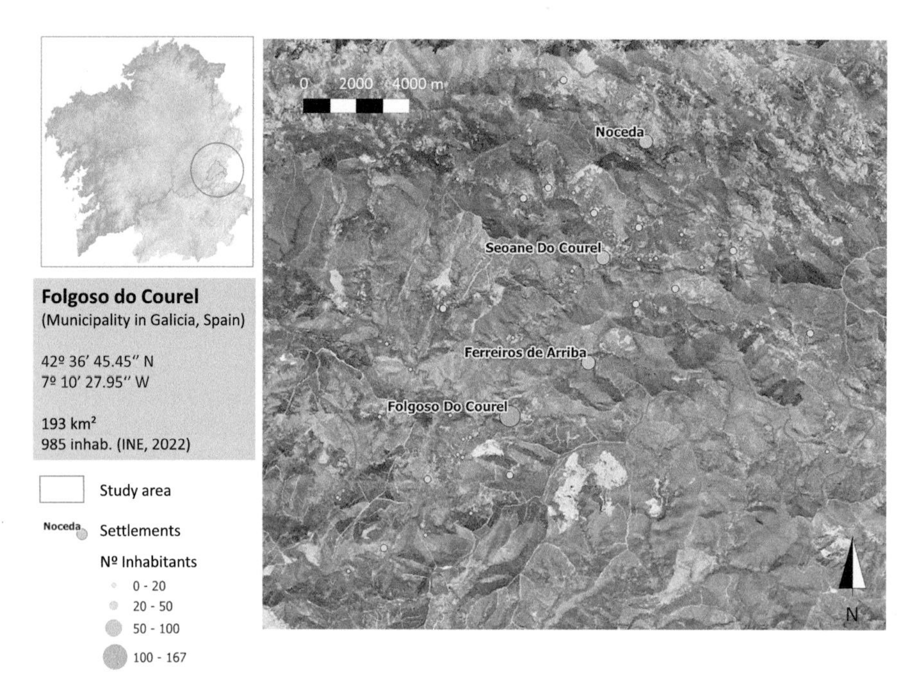

Fig. 3.2 Study area: local level. (Source: Prepared by authors; Aerial imagery from Centro Nacional de Información Geográfica (2019); Demographic data from INE (2022))

Table 3.1 Basic information of the study area

Country	Spain
Province	Lugo
District	O Courel
Municipality	Folgoso do Courel
Size of geographical area—Galicia—(hectare)	2,957,700
Dominant ethnicity(ies), if appropriate	Caucasic
Size of case study/project area (hectare)	19,300
Dominant ethnicity in the project area	Caucasic
Number of direct beneficiaries (people)	985
Number of indirect beneficiaries (people)	2,691,456
Geographic coordinates (latitude, longitude)	42°36'45.45" N 7°10'27.95" W

Source: Prepared by authors, Data from INE (2022)

the intellectual property rights of the territory and of all necessary information about value-addition/end-products for consumers.

2. Then, we used a Geographical Information System (GIS) to precisely locate chestnut production areas and analyze land uses. Also, spatial data of recent forest wildfires was overlaid to quantify the impacts of fires (MITECO 2011; Consellería de Medio Rural 2022a).

3. Finally, we completed our analysis by interviewing business owners to elicit business-SEPL relationships, problems, and opportunities for *souto*-dependent businesses in the area. In an exploratory approach, we selected local businesses that satisfied the characteristics of (a) being in some way dependent on the local production of chestnut and (b) developing local chestnut-related products. Two businesses that met these characteristics were interviewed. The approach for the interviews was qualitative and employed semi-structured open discussions covering:

   - Orientation of production/marketable products
   - Relationships with producers/suppliers of raw material (i.e. chestnuts harvested in the *soutos*)
   - Customer characteristics and spatial scope and targeting of the products
   - Other factors aside from chestnut fruit production and/or use
   - Associations and networks, including economic, social, and cultural
   - Biodiversity and landscape, and perceived relation with their businesses

3 Results

3.1 General Socio-Economic Features

Fologoso do Courel is a municipality with a clear decreasing trend in its population. Figure 3.3 shows the demographic trend from the year 1877. The decline can be traced back to the 1910s, but a clear change in the slope can be identified from 1940.

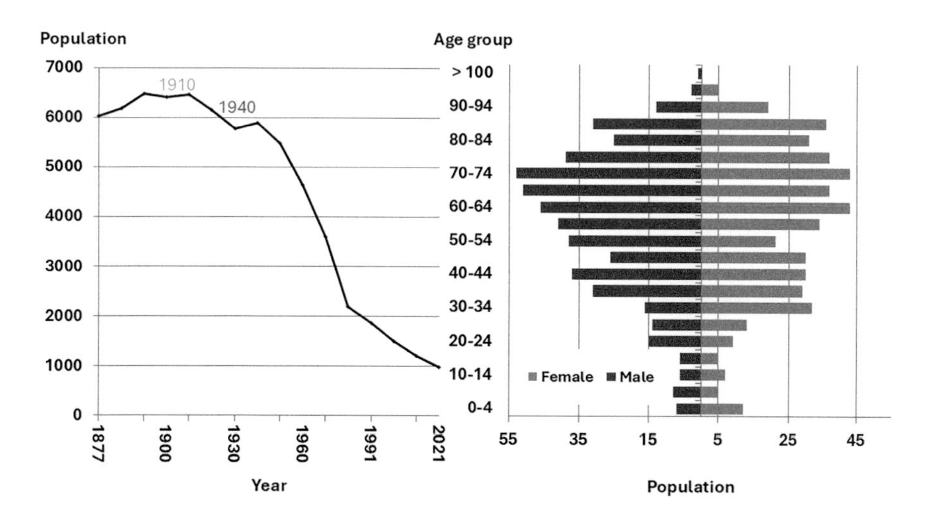

Fig. 3.3 Left, population trend in Folgoso do Courel, in total number of inhabitants since 1877, with vertical axis showing population and horizontal axis showing years; Right, population pyramid with values from 1 January 2022, with vertical axis showing age groups and horizontal axis showing number of inhabitants. (Source: Prepared by authors; Data from INE (2022))

The current situation, with 985 inhabitants, shows a population growth rate of −22 (INE 2022), so generational replacement is not warranted.

This demographic situation is also reflected in the occupational structure: 43.2% of the population is retired, 12.6% unemployed, and 7% in other labour situations (e.g. working in non-economic activities or unable to work), and 37% is employed. Around 26% of employed people is engaged in agriculture, 15% in industry, and 53% in the service sector. While statistics show an important shift towards the service sector of the economy, we have to take into account that many retired people are still working in agriculture as a means of supplementing their income.

3.2 General Biophysical and Bioeconomic Features

Chestnut forests, including *soutos* in different stages of management (i.e. abandoned, in different stages of abandonment, or fully functional and productive; see Roces-Diaz et al. 2018), and new plantations span 2756 hectares in Folgoso do Courel (MITECO 2011), implying approximately 14.3% of the municipality area. Other important land cover types are native forests made up of one dominant species (12.1%) or mixed species (12.2%), multi-species fast-growing forest (1.1%), and single-species fast-growing forest (11.8%). Non-forestry areas, usually agricultural land and shrubland, span 48.42% of the total surface.

Analysis of land surface affected by forest fires in 2022 (See "Burnt area" overlaid with other land cover classes in Fig. 3.4) showed that most of the burnt forested areas correspond to exotic single-species (67%) or multi-species (84%) fast-growing forest stands, while native forests showed relatively low impacts in one dominant species stands (9%) or multi-species mixed (13.2%), with chestnut forests affected at a rate of 25.5% (Percentages are the proportion of the total of each land cover type prior to the wildfire.)

3.3 Specific Examples of Local Businesses and Business Structure

3.3.1 Protected Geographical Indication

The interviews with members of the Regulation Council of the Protected Geographical Indication "Galician Chestnut" (see Sect. 2.2) allowed for identification of the main typologies of actors and their relationships. This information was a valuable contribution to the development of a basic representation of the chestnut SEPL focused on the socio-economic subsystem (see details in discussion and Fig. 3.6).

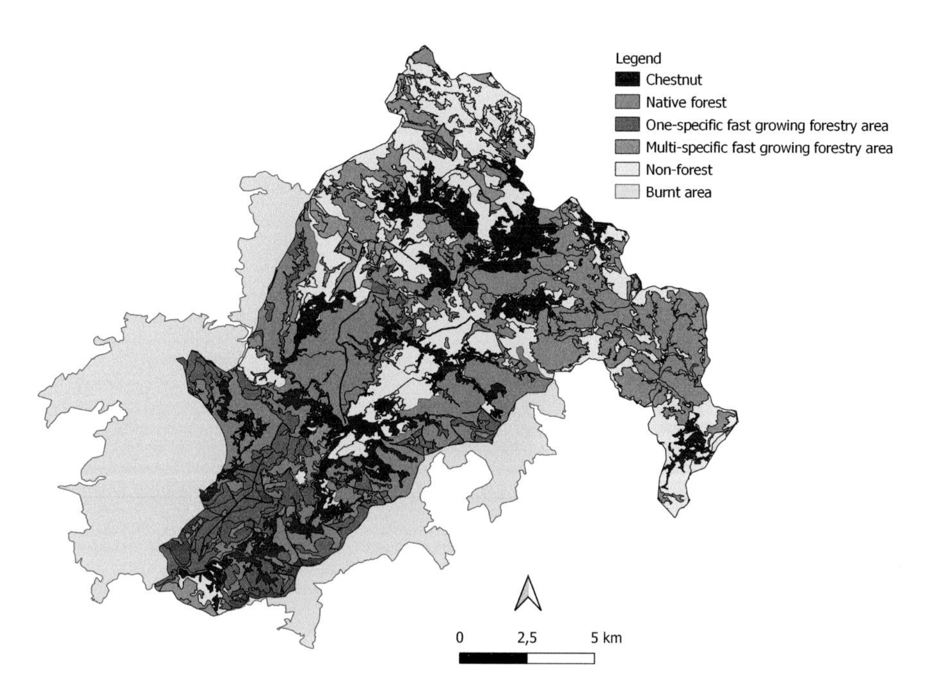

Fig. 3.4 Chestnut forest distribution in the study area (in dark green). Grey shadow shows the area affected by forest fires in 2022. (Source: Prepared by authors; Land cover data from MITECO (2011); Wildfire data from Consellería de Medio Rural (2022a))

3.3.2 Study Case 1: Diversified Food Processing Business

The first interviewee is native from Courel who started a trading business. Together with his wife in 2008, he founded a business based on the elaboration of different food products in the nearby municipality of Quiroga. Product diversification became necessary due to constraints related to the COVID-19 crisis that make it difficult to make a living just from chestnut production. The business uses chestnut fruit as raw material for an estimated 40% of their products, producing jams, syrups, compotes, or just preserved fruits. Habitual customers are consumers reached directly or through markets at local fairs or, especially when customers are already familiar with products, through the Internet. Initially, the family planned to sell raw chestnut fruits but later started making valued-added products when they realized that the margin of profit increases by 400% with value addition. The interviewee identified the low prices of chestnuts paid by intermediaries to producers and the ageing population as the most important factors driving abandonment of chestnut-based activity in the area.

Nevertheless, the interviewee emphasized the importance of plant and fruit variety. From the economic point of view, price levels and their evolution are also dependent on quality, which is directly related to the fruit varieties and their

Fig. 3.5 Products made from chestnuts (cookies and jams) and honey from *soutos* in Courel. (Photo by authors)

diversity. He named important varieties like *branca*, *verdea*, *parede*, and *presa*. The reasons for this importance include differences in seasonality and the possibility of a continuous supply if an adequate varietal diversity is found in the *soutos*, as well as suitability to different uses depending on the characteristics of the varieties.

The business's link with the territory was prominent, as not only do they have their own *soutos* as well as individual trees for chestnut production, but also use mushrooms, blackberries, and blueberries growing wild in the *soutos* for preparations. In the past, they also produced honey (Fig. 3.5). The interviewee spoke with strong conviction about the relationship between the quality of their products and the place of origin:

> The chestnut fruits (…) of the Courel area are the best in the world (…) Starting from this, evidently you must value what you have: chestnuts from very nice varieties, a spectacular place, an incredible tradition and culture around the chestnut. Then, you must sell that.

For this business, the connection to the *souto* is multi-level and acknowledges cultural elements as important, especially the landscape and traditional knowledge. Regarding the latter, the interviewee recognized that much traditional knowledge on management (e.g. planting, pruning or selecting trees, and caring for the whole *souto*) has already been lost. As he stated:

> (…) the *souto* is lost when the elder people are gone.

This aspect of multifunctionality was also underscored in comments on the impacts of recent wildfires, that in addition to highlighting the protective function of *soutos* (as they are less flammable than fast-growing species), warned about the proliferation of fast-growing species around settlements:

> In the Courel (...) area, all villages are surrounded by *soutos* (...) the recent wildfires confirmed that villages are surrounded by chestnut trees for a reason – not only because they produce fruit – if a wildfire occurs, chestnut trees don't burn (...) if you surround a village with fuel, in the end it will burnout.

During the last few years (especially in the last decade), chestnuts have suffered impacts from plagues (e.g. the chestnut gall wasp, *Dryocosmus kuriphyllus*), diseases (e.g. those caused by fungi *Phytophthora cinnamomi* and *Cryphonectria parasitica*), and droughts in summer that together caused a major decrease in production. As these impacts, together with the ageing population, are generalized in the area, it was difficult for the business to find local product sellers that could compensate for the decreased production in their own *soutos*. Consequently, they had to adapt their level of production to the lack of chestnut fruit.

3.3.3 Study Case 2: Chestnut-Specialized Food-Processing Business

The second example is a business in Folgoso do Courel, founded by two women. They started making chestnut products when they observed a demand for chestnut-related products at local cultural events but a lack of supply. They started producing chestnut cakes, but soon extended their production to chestnut cookies and preserved fruits. The products were bought by individuals as well as restaurants. They noted the importance of coordinating with other producers through informal networks (e.g. sharing resources in packaging and attending markets). They have received some special requests. In particular, last Christmas they produced 5000 biscuit bags for a big company from Madrid that were integrated in gift boxes planned for local products from areas impacted by wildfires.

Nowadays, they feel rather pessimistic about the continuity of the business as a result of ageing and the increasing frequency of summer droughts, pests, and diseases. Government agencies do assess damages and propose solutions, but the formalities required to take advantage of these are too complicated. These problems led them to drastically reduce/abandon their *soutos* and buy around 1000 kg of dried chestnut fruits/chestnut flour from nearby *soutos* to meet their needs.

In any case, they highlighted their dependence on chestnuts from *soutos*, as opposed to industrially produced ones:

> We don't buy industrially-produced chestnut fruits, ours are all produced (...) in *soutos* [and we see it as interesting that] we keep doing like that. Could you imagine bringing in industrialized chestnuts?

As in the former case, the values linked to the *soutos* are perceived as multiple. While economic activity is apparently driven mostly by instrumental values—those related to the obtainment of direct benefits from the chestnut forest ecosystem—it is apparent that relational values—those reflecting a symbolic dimension and a sense of identity in the relationships with nature—were also behind the motivation to develop their business. In addition to the interest in preserving the diversity of chestnut varieties and the use of chestnuts produced in *soutos*, there is also a clear interest in the preservation of related cultural traditions and landscapes. This was especially evident when asked about the importance of *soutos* in the local landscape:

> [*Soutos* integration in the landscape is] vital. Because it is culture, it is tradition, and it is landscape. Imagine taking out the *soutos* [from the landscape]. And here many are decaying. But here all the villages have their *soutos*.

Also, they considered their activities' contribution to biodiversity and landscape conservation through the culture of *soutos* to be very important.

4 Discussion

Integration of the information gathered from interviews and field observations led us to a graphic outline of the socio-economic subsystem of the chestnut SEPL (Fig. 3.6).

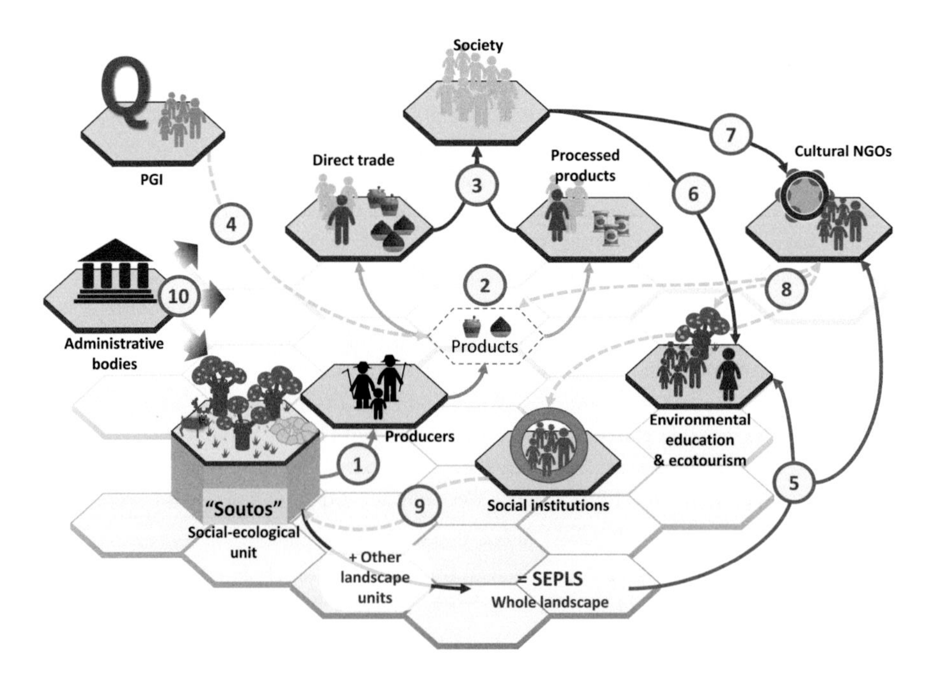

Fig. 3.6 Synthetic scheme for the integration of business activities in the SEPL, each line-patterned hexagon symbolizing an interacting element in the socio-economic sub-system. See text for details. (Source: Prepared by authors)

The starting point in the figure are the *soutos* as the socio-ecological units in the SEPL system. Their management allows producers to obtain products like chestnut and honey (1), and also to benefit from other provisioning (e.g. wood), regulating (e.g. protection against fire), and cultural (e.g. traditional knowledge) ecosystem services. These products are integrated in the market through direct trade, or as raw material for value-added products (2). Different profiles of final customers access the outcoming products by buying them at the local, regional, and even national level (3). The intervention of the Protected Geographical Indication (PGI) organization enhances the value of products and credibility of producers as well, and supports coordination between producers, processors, and other actors in the chestnut sector (4). On the other hand, *soutos* are an indivisible part of the aesthetical and cultural values of the whole landscape that are reflected in both local environmental education/eco-tourism businesses and cultural NGOs that preserve traditional knowledge (5) and promote activities with a diverse demand from society (6, 7). Specifically, activities by cultural NGOs sustain important linkages with environmental education, promotion of local products, and visibility of social institutions (8), that in turn provide essential support for the *soutos* (9). Finally, administrative bodies (10) have the responsibility to directly support all stakeholders and the functions and integrity of the SEPL.

Nevertheless, relationships between economic activity and ecosystems in the study area cannot be simplified into linear or direct relationships. This should be expected, as socio-ecological systems are by definition complex adaptive systems (Levin et al. 2013; Preiser et al. 2018) often exhibiting surprising or unexpected effects. As mentioned, interrelationships between economic activities and biodiversity may be represented by *risks*, *impacts*, and *opportunities*.

For instance, *soutos*-related businesses suffer from *risks,* both of a biophysical (e.g. pests, diseases, and wildfires) and economic (e.g. uncertainties in prices and availability of suppliers) nature. Wildfires heavily affected the area in 2022 (see Sect. 3.2 and Fig. 3.7) with immediate, destructive effects. While wildfires are not a

Fig. 3.7 Post-fire image of a *souto* in the Vilar settlement, affected by the 2022 wildfires. (Photo credit: Orlando Gregorio Álvarez-Álvarez, neighbour; reproduced with permission)

new phenomenon, their underlying causes and spatial patterns vary in space and time (Bowman et al. 2011). Due to climate change (Lindner et al. 2010; Sousa-Silva et al. 2018) and the increasing connectivity and extension of vegetation with high fuel load capacity (Castellnou and Miralles 2009; Moreira et al. 2020), wildfires are becoming progressively more relevant today. There has been an increase in the occurrence, extent, and severity of forest fires. At a finer scale, vulnerability may depend on landscape pattern (Calviño-Cancela et al. 2017), with mosaic areas being less vulnerable than extensive areas of fast-growing afforested species or some types of shrubland. The results of our land cover analysis provide support for this difference in vulnerability (see Fig. 3.4 and related text). Thus, landscapes including well-managed chestnut *soutos* offer the *opportunity* for *fire-smart* management (Pais et al. 2020).

On the other hand, the *risks* of pests and diseases often cause fragmentation of chestnut orchards apart from decline in economic yields. Specifically, for the recently arrived exotic pest *Dryocosmus kuriphyllus*, access to programmes by the regional administration aimed at biological control using the exotic parasitoid *Torymus sinensis* (Nieves-Aldrey et al. 2019) can be hindered by administrative barriers and lack of communication, as reported by the local population. Additionally, the use of an exotic species for biological control should be monitored to avoid impacts on local biodiversity (Gil-Tapetado et al. 2023). At longer temporal scales, the capacity of the SEPL for the conservation of a high genetic diversity in chestnut trees could be a crucial opportunity, noting the observed differences in susceptibility to *Dryocosmus kuriphilus* by chestnut tree genotypes (Lombardero et al. 2022; Castedo-Dorado et al. 2023). As mentioned, such genetic diversity also has economic importance for the diversification of fruit-related products.

However, as interviewees stated, economic benefits from chestnuts have dramatically declined because of the combined/synergistic effects of climate change, demographic change, pest/pathogen infestation, and habitat homogenization from deagrarianization over the last few decades (López-Iglesias 2006; López-Iglesias et al. 2013). Courel is one of the areas in Europe suffering the greatest *impacts* from loss of utilized agricultural area (UAA) (Murua et al. 2013). These factors combined put at *risk* the continuity of the local supply of chestnut fruit. Abandonment of traditional activities can be, from the socio-economic point of view, associated with the demographic decline, a continuous process since the beginning of the last century that accelerated from the 1940 and 1950s. Causes for these demographic dynamics are multiple, and some of them correlate with the socio-ecological transitions associated with social, institutional, and technological changes (Guzmán-Casado et al. 2018; González de Molina et al. 2019). Nevertheless, two demographic components should be noted: (i) decline in numbers is, in itself, due mainly to successive waves of migration of younger people from rural areas to cities (Dubert et al. 2019), in a great extent associated with the deagrarianization process (Pérez-Fra et al. 2006), and (ii) the ageing population is consequence of both the absence of the migrated young population and increased lifespan. This begs policy interventions to maintain *soutos* and the *soutos*-based economy to create new social and economic *opportunities*. The ageing population, in fact, allows for conservation of traditional knowledge but not its practice, more so for labour intensive practices, in youth outmigration scenarios; also, it challenges the capacity of rural areas to provide social services. In any case, deagrarianization and associated abandonment can have important direct and indirect *impacts* on *soutos*-related businesses through interactions with other ecosystems and economic activities in the landscape. As previously mentioned, one possible outcome is the progressive substitution of *soutos* with fast-growing forest plantations or natural vegetation arising from ecological succession. The aforementioned changes are likely to increase wildfire vulnerability and thus may negatively *impact* ecological (Deus et al. 2018) as well as social systems (Martínez-Cabrera et al. 2020). Nonetheless, natural succession in abandoned orchards (see e.g. Díaz-Varela et al. 2011; Roces-Diaz et al. 2018) could be an *opportunity* for enhancement of biodiversity including umbrella species such as the brown bear (*Ursus arctos*) (Pérez-Girón et al. 2022).

Lastly, it is important to note that sustainable economic sectors and activities, such as the ones that are the focus of this study, have a strong potential to contribute to a knowledge transmission process based on the co-production of knowledge involving different sectors of society (Ruckelshaus et al. 2015; Posner et al. 2016). The activities themselves could help to reveal the importance of biodiversity and ecosystem services for society, facilitate the development of a common language, and generate actions aimed at the production of specific results in the preservation and maintenance of ecosystems.

5 Conclusions

The persistence of economic activities related to *soutos* presents multifaceted opportunities for the communities and ecosystems of the Courel mountains. Businesses created around the processing of chestnut fruits and other products such as honey, berries, or mushrooms are closely linked to the continuity of the *soutos*, not only in their productive aspect but also as part of the natural capital and cultural heritage of the area. Consequently, economic activities related to the management of *soutos* and their related socio-ecological production landscapes play an important role in nature's contributions to people, including important flows of provisioning, regulating, and cultural ecosystem services. Their supply is dependent on ecological integrity and ecosystem functioning, which is also reflected in the concurrence of relational and intrinsic values in the perspectives and views of local actors. On the other hand, the maintenance of these forms of interaction between society and nature has a strong potential to preserve biodiversity at the genetic, species, and ecosystem levels, with direct and indirect linkages to more intrinsic values of landscape and nature. This constitutes the basis for the extension of the described multifunctional character to other economic activities dependent on the mountain landscapes where the *soutos* are an essential element. These include environmental education, agri-tourism, and eco-tourism, for which institutional initiatives related to rural development could provide essential support. Also, institutional level responses, such as the recently approved Strategic Program for Chestnut at the regional level and the existing classification of *soutos* as "Castanea sativa woods," an Annex I habitat type (code 9260) in the European Union's Habitat Directive, could be potentially useful to strengthen the *soutos*, provided they are appropriately implemented.

Nevertheless, the future of *soutos* as a multifunctional element with high capacity for biodiversity conservation may be hindered by risks and impacts of a biophysical and socio-economic nature, showing a complex interlinkage between causes and effects and elements and connections in the socio-ecological system. The involvement of different actors, institutions, and administration directly or indirectly related to the SEPL will be necessary for the co-production of knowledge and the development of pluralistic, holistic views for integrated ways of management and restoration of the socio-ecological chestnut production landscapes.

References

Aguín O, Rial C, Piñón P, Sainz MJ, Mansilla JP, Salinero, C (2023) First Report of *Gnomoniopsis smithogilvyi* Causing Chestnut Brown Rot on Nuts and Burrs of Sweet Chestnut in Spain. Plant Disease 107(1):218

Barral MP, Benayas JMR, Meli P, Maceira NO (2015) Quantifying the impacts of ecological restoration on biodiversity and ecosystem services in agroecosystems: A global meta-analysis. Agric Ecosyst Environ 202:223–231

Bowman DMJS, Balch J, Artaxo P, Bond WJ, Cochrane MA, D'Antonio CM, DeFries R, Johnston FH, Keeley JE, Krawchuk MA, Kull CA, Mack M, Moritz MA, Pyne S, Roos CI, Scott AC, Sodhi NS, Swetnam TW (2011) The human dimension of fire regimes on Earth. Journal of Biogeography 38(12):2223–2236

Calviño-Cancela M, Chas-Amil ML, García-Martínez ED, Touza J (2017) Interacting effects of topography, vegetation, human activities and wildland-urban interfaces on wildfire ignition risk. For. Ecol. Manag. 397:10–17

Castedo-Dorado F, Álvarez-Álvarez P, Cuenca Valera B, Lombardero MJ (2023) Local-scale dispersal patterns and susceptibility to Dryocosmus kuriphilus in different Castanea species and hybrid clones: Insights from a field trial. New Forests 54(1):9–28

Castellnou M, Miralles M (2009) The changing face of wildfires. Crisis Response 5(4):56–57

Centro Nacional de Información Geográfica (2019) Centro de Descargas, Madrid, viewed 25 February 2023. Retrieved from http://centrodedescargas.cnig.es/CentroDescargas/index.jsp

Commission Regulation (EU) No 409/2010, of 11 May 2010, entering a name in the register of protected designations of origin and protected geographical indications (Castaña de Galicia (PGI))

Consellería de Medio Rural (2021) *IXP* Castaña de Galicia, Galician Ministry for Rural Areas, Xunta de Galicia, Santiago de Compostela, viewed 25 February 2023. Retrieved from https://mediorural.xunta.gal/es/temas/alimentacion/productos-gallegos-de-calidad/productos-origen-vegetal/castana-galicia

Consellería de Medio Rural (2022a) Perimeter of wildfire impacting Courel Mountain Areas, Dirección Xeral de Defensa do Monte, Galician Ministry for Rural Areas, Xunta de Galicia, Santiago de Compostela, unpublished data

Consellería de Medio Rural (2022b) Programa Estratéxico do Castiñeiro, Grupo de investigación G4 Plus, Universidade de Vigo-Dirección Xeral de Planificación e Ordenación Forestal, Galician Ministry for Rural Areas, Xunta de Galicia, Santiago de Compostela

Deus E, Silva JS, Castro-Díez P, Lomba A, Ortiz ML, Vicente J (2018) Current and future conflicts between eucalypt plantations and high biodiversity areas in the Iberian Peninsula. J. Nat. Conserv. 45:107–117

Díaz-Varela RA, Álvarez-Álvarez P, Diaz-Varela E, Calvo-Iglesias S (2011) Prediction of stand quality characteristics in sweet chestnut forests in NW Spain by combining terrain attributes, spectral textural features and landscape metrics. For. Ecol. Manag 261(11):1962–1972

Dubert I, Cernadas XMA, Castelao OR, Romero HM, Villares R, García FG, Caramés AP, Pardo AL, Vázquez RS, Borge JH, Táboas DL, Fernández MF (2019) A morte de Galicia, Edicións Xerais, Santiago de Compostela

European Environment Agency (2017) Biogeographical Regions in Europe, Copenhagen, viewed 25 February 2023. Retrieved from https://www.eea.europa.eu/data-and-maps/figures/biogeographical-regions-in-europe-2

Fernández-López J, Fernández-Cruz J, Míguez-Soto B (2021) The demographic history of Castanea sativa Mill. in southwest Europe: A natural population structure modified by translocations. Mol. Ecol. 30:3930–3947

Garbach K, Milder JC, Montenegro M, Karp DS, DeClerck FAJ (2014) Biodiversity and ecosystem services in agroecosystems. Encyclopedia of agriculture and food systems 2:21–40

Gil-Tapetado D, López-Estrada EK, Jiménez Ruiz Y, Cabrero-Sañudo FJ, Gómez JF, Durán Montes P, Rey del Castillo C, Rodríguez-Rojo MP, Polidori C, Nieves-Aldrey JL (2023) Torymus sinensis against the invasive chestnut gall wasp: Evaluating the physiological host range and hybridization risks of a classical biological control agent. Biological Control 180:105187

González de Molina M, Soto Fernández D, Guzmán-Casado G, Infante-Amate J, Aguilera Fernández E, Vila Traver J, García Ruiz R (Eds.) (2019) Historia de la agricultura española desde una perspectiva biofísica, 1900-2010, Gobierno de España, Ministerio de Agricultura, Pesca y Alimentación, Madrid

Guzmán-Casado GI, Aguilera E, García-Ruiz R, Torremocha E, Soto-Fernández D, Infante-Amate J, González de Molina M (2018) The agrarian metabolism as a tool for assessing agrarian sustainability, and its application to Spanish agriculture (1960-2008). Ecol. Soc. 23(1):2

Hanson C, Finisdore J, Ranganathan J, Iceland C (2013) The Corporate Ecosystem Services Review: Guidelines for Identifying Business Risks and Opportunities Arising from Ecosystem Change, Version 2.0, World Resources Institute, Washington, DC

IGE (2023) Instituto Galego de Estatística (Galician Institute for Statistics), viewed 02 February 2023. Retrieved from www.ige.gal

INE (2022) *Instituto Nacional de Estadística (Spanish Statistical Office) – Nomenclátor, Nomenclátor: Población del Padrón Continuo por unidad poblacional*, viewed 02 February 2023. Retrieved from http://bit.ly/2WToH7L

Levin S, Xepapadeas T, Crépin AS, Norberg J, Zeeuw A, Folke C, Hughes T, Arrow K, Barrett S, Daily G, Ehrlich P, Kautsky N, Mäler KG, Polasky S, Troell M, Vincent JR, Walker B (2013) Social-ecological systems as complex adaptive systems: Modeling and policy implications. Environ Dev Econ 18(2):111–132

Lindner M, Maroschek M, Netherer S, Kremer A, Barbati A, Garcia-Gonzalo J, Seidl R, Delzon S, Corona P, Kolström M, Lexer MJ, Marchetti M (2010) Climate change impacts, adaptive capacity, and vulnerability of European forest ecosystems. For. Ecol. Manag. 259(4):698–709

Lombardero MJ, Ayres MP, Álvarez-Álvarez P, Castedo-Dorado F (2022) Defensive patterns of chestnut genotypes (Castanea spp.) against the gall wasp, Dryocosmus kuriphilus. Front. for. glob. change 5:1046606

López-Iglesias E (2006) Axuste agrario e despoboamento rural en Galiza: análise das tendencias recentes e reflexións cara o futuro. In: Foro Rural Galego (ed) *Congreso Técnico-Científico sobre Desagrarización e Sostibilidade Rural na Euro-Rexión Galicia-Norte de Portugal*, Fondo Rural Galego, Lugo, pp 89–100

López-Iglesias E, Sineiro-García F, Lorenzana-Fernández R (2013) Processes of farmland abandonment: land use change and structural adjustment in Galicia (Spain). In: Ortiz-Miranda D, Moragues-Faus A, Arnalte-Alegre E (eds) Agriculture in Mediterranean Europe: between old and new paradigms. Emerald Group Publishing Limited, Bingley, pp 91–120

MAPA (2021) *Ministerio de Agricultura, Pesca e Alimentación - Anuario de Estadística 2021*, Ministry of Agriculture, Fisheries and Food, Madrid, viewed 25 February 2023. Retrieved from https://www.mapa.gob.es/es/estadistica/temas/publicaciones/anuario-de-estadistica/2021/default.aspx

MAPA (2022) *Ministerio de Agricultura, Pesca e Alimentación - Análisis y prospectiva-Informe interactivo de comercio exterior*, Ministry of Agriculture, Fisheries and Food, Madrid, viewed 25 February 2023. Retrieved from https://www.mapa.gob.es/es/ministerio/servicios/analisis-y-prospectiva/powerbi-comex.aspx

Martínez-Cabrera H, Rodríguez G, Ballesteros H (2020) Degradation of social institutions and land use: unfolding feedback mechanisms between afforestation with fast-growing species and living conditions in rural areas. Revista Galega de Economía 29(2):1–18

MITECO (2011) *Ministerio para la transición ecológica y el reto demográfico-Cuarto Inventario Forestal Nacional: Galicia.* Ministry for Ecological Transition and Demographic Challenge, Madrid, viewed 25 February 2023. Retrieved from https://www.miteco.gob.es/es/biodiversidad/temas/inventarios-nacionales/inventario-forestal-nacional/cuarto_inventario.aspx

Moreira F, Ascoli D, Safford H, Adams MA, Moreno JM, Pereira JMC, Catry FX, Armesto J, Bond W, González ME, Curt T, Koutsias N, McCaw L, Price O, Pausas JG, Rigolot E, Stephens S, Tavsanoglu C, Vallejo VR, Van Wilgen BW, Xanthopoulos G, Fernandes PM (2020) Wildfire management in Mediterranean-type regions: Paradigm change needed. Environ. Res. Lett. 15(1):011001

Murua JR, Astorkiza I, Eguía B (2013) Conflicts between Agricultural Policy and Sustainable Land Use: The Case of Northern Spain. Panoeconomicus 3:397–414

Nieves-Aldrey JL, Gil-Tapetado D, Gavira O, Boyero JR, Polidori C, Lombardero MJ, Blanco D, Rey del Castillo C, Rodriguez Rojo P, Vela JM, Wong E (2019) Torymus sinensis Kamijo a biocontrol agent against the invasive chestnut gall wasp Dryocosmus kuriphilus Yasumatsu in Spain: its natural dispersal from France and the first data on establishment after experimental releases. Forest Systems 28(1):e001

Pais S, Aquilué N, Campos J, Sil Â, Marcos B, Martínez-Freiría F, Domínguez J, Brotons L, Honrado JP, Regos A (2020) Mountain farmland protection and fire-smart management jointly reduce fire hazard and enhance biodiversity and carbon sequestration. Ecosyst. Serv. 44:101143

Pereira-Lorenzo S, Ramos-Cabrer AM, Barreneche T, Mattioni C, Villani F, Díaz-Hernández B, Martín LM, Robles-Loma A, Cáceres Y, Martín A (2019) Instant domestication process of European chestnut cultivars. Ann. Appl. Biol. 174:74–85

Pérez-Fra M, García-Arias AI, Docio Rodriguez F (2006) Efectos territoriales de la reestructuración de la ganadería bovina de la Cornisa Cantábrica. In: Arnalte Alegre E (ed) Políticas agrarias y ajuste estructural en la agricultura española, Ministerio de Agricultura, Pesca y Alimentación, Madrid, pp 327–350

Pérez-Girón JC, Díaz-Varela ER, Álvarez-Álvarez P, Hernández Palacios O, Ballesteros F, López-Bao JV (2022) Linking landscape structure and vegetation productivity with nut consumption by the Cantabrian brown bear during hyperphagia. Sci. Total Environ. 813:152610

Posner SM, McKenzie E, Ricketts TH (2016) Policy impacts of ecosystem services knowledge. Proc. Natl. Acad. Sci. U.S.A. 113:1760–1765

Preiser R, Biggs R, De Vos A, Folke C (2018) Social-ecological systems as complex adaptive systems: Organizing principles for advancing research methods and approaches. Ecol. Soc. 23(4):46

Regulation (EU) No 1151/2012 of the European Parliament and of the Council of 21 November 2012 on quality schemes for agricultural products and foodstuffs

Roces-Diaz JV, Díaz-Varela ER, Barrio-Anta M, Álvarez-Álvarez P (2018) Sweet chestnut agroforestry systems in North-western Spain: Classification, spatial distribution and an ecosystem services assessment. For Syst 27(1):03

Ruckelshaus M, McKenzie E, Tallis H, Guerry A, Daily G, Kareiva P, Polasky S, Ricketts T, Bhagabati N, Wood SA, Bernhardt J (2015) Notes from the field: Lessons learned from using ecosystem service approaches to inform real-world decisions. Ecol Econ 115:11–21.

Sousa-Silva R, Verbist B, Lomba Â, Valent P, Suškevičs M, Picard O, Hoogstra-Klein MA, Cosofret VC, Bouriaud L, Ponette Q, Verheyen K, Muys B (2018) Adapting forest management to climate change in Europe: Linking perceptions to adaptive responses. For Policy Econ 90:22–30

WBCSD (2011) Guide to Corporate Ecosystem Valuation. World Business Council on Sustainable Development, Geneva, viewed 17 February 2023. Retrieved from https://www.wbcsd.org/contentwbc/download/573/6341

Zhang W, Gowdy J, Bassi A, Santamaria M, Declerck F, Adegboyega A, Andersson G, Augustyn AM, Bawden R, Bell A, Darnhofer I, Dearing J, Dyke J, Failler P, Galetto L, Hernández C, Johnson P, Jones S, Kleppel G, Komarek AM, Latawiec A, Mateus R, McVittie A, Ortega E, Phelps D, Ringler C, Sangha KK, Schaafsma M, Scherr S, Hossain MS, Thorn JPR, Tyack N, Vaessen T, Viglizzo E, Walker D, Willemen L, Wood SLR (2018) Systems thinking: an approach for understanding 'eco-agri-food systems'. In: TEEB (ed) TEEB for Agriculture & Food: Scientifc and Economic Foundations, UN Environment, Geneva, pp 17–55

Emilio R. Díaz-Varela Associate Professor at the University of Santiago de Compostela, Spain. His research interests are social-ecological systems analysis and management, spatial analysis in complex patterns, and sustainable development and resilience in urban and rural systems.

Pedro Álvarez-Álvarez Associate Professor in the Department of Organisms and Systems Biology, University of Oviedo, Spain. Pedro is interested in sustainable forest management, complex forests, and traditional chestnut (Castanea sativa Mill.) stands.

José C. Pérez-Girón Postdoctoral researcher at the Department of Botany of the University of Granada, Spain. He is interested in primary production indicators and the potential effects of climate change on plant and animal species.

Ramón A. Díaz-Varela Associate Professor at the University of Santiago de Compostela, Spain. Research interests are the use of geomatics for the assessment of biodiversity at different scales.

María A. Ferreira Golpe Researcher in the Department of Applied Economics at the University of Santiago de Compostela, Spain. Works on projects focused on beekeeping, invasive species, and rural policy.

Ana I. García-Arias Associate Professor in the Department of Applied Economics at the University of Santiago de Compostela, Spain. Her publications are related to agricultural economics, the evaluation of agri-environmental policies, and the identification of ecosystem services.

Chapter 4
Traditional Environmental Knowledge and Trees Conservation: The Example of the White Truffle (*Tuber Magnatum Pico*) in Italy

Guido Gualandi

Abstract Truffles are edible ectomycorrhizal fungi that grow mostly in temperate areas of Mediterranean Europe, western North America, South Africa, and Australia, though they are most commonly harvested in Italy, France, and Spain. Certain types of truffles, such as the white truffle, grow mainly in the wild. The truffle industry worldwide generates hundreds of millions of euros annually and positively impacts urban and rural areas in Italy by increasing tourism and hospitality. It also contributes to maintaining biodiversity in wild areas and to sustaining traditional land management practices. However, the harvest of white truffles (*Tuber magnatum Pico*) specifically has been declining for several decades due to climate change and changing conditions such as reduced tree biodiversity, poor soil and water management practices, and industrial agricultural methods. Social changes and loss of Traditional Environmental Knowledge (TEK) have also had a negative impact on the harvesting of white truffles.

This research was conducted in Piedmont, Emilia-Romagna, and Tuscany (Italy) through conducting interviews with truffle hunters and harvesters and gathering data from local stakeholders.

The habitat of the white truffle is endangered by human activities, the reduction of forested areas, and climate change. Forests or groups of trees where truffles grow need to be kept and replanted properly with respect for the biodiversity that supports the truffle habitat. Diminishing shady areas accelerates heating and drying of the soil. The reduction of forested area also changes the microclimate and reduces moisture, starting a chain reaction leading to further reduction in truffle biodiversity. Poor management of waterways also destroys areas for truffle growth. The largest impact is on wild varieties such as *Tuber magnatum Pico*, the white truffle.

All the photos in this work belong to Associazione nazionale Città del tartufo.
All the tables and graphics were made by the author.

G. Gualandi (✉)
Comunità dei Grani antichi, Montespertoli, Italy
e-mail: guido@guidogualandi.com

M. Nishi et al. (eds.), *Business and Biodiversity*, Satoyama Initiative Thematic Review, https://doi.org/10.1007/978-981-97-7574-3_4

Keywords Traditional local knowledge · Traditional environmental knowledge (TEK) · Payment for ecosystem services (PES) · White truffle · Tuber magnatum Pico · Mycorrhiza · Fungus · Italy · Biodiversity

1 The White Truffle in Italy

1.1 Definition of the Area and Scope of the Study

Truffles are edible fungi that grow mostly in temperate areas of Mediterranean Europe, western North America, South Africa, and Australia, though traditionally they are mostly harvested in Italy, France, and Spain. Truffles form large underground fruit bodies. They live in ectomycorrhizal association with a broad variety of hosts (trees) in a variety of habitats such as temperate forests, boreal forests, floodplains, tree nurseries, restoration sites, and Mediterranean woodlands. They reproduce through spores. The most precious edible fungi are types of scented truffles: *Tuber magnatum Pico*—the white truffle—and *Tuber melanosporum Vittad*—the Perigord black truffle. These species are mainly limited to the Mediterranean region. There are many other truffle species that are edible and sold.

This study will focus on the white truffle (*Tuber magnatum Pico*) as its habitat is the most endangered by the loss of biodiversity and climate change. The white truffle is not cultivated, and conservation of its habitat and its collection rely heavily on Traditional Environmental Knowledge (TEK). TEK is also an important component of the local economy and cultural heritage. Since 2021, truffle hunting has been designated by UNESCO as Intangible Cultural Heritage, recognizing the role of the *tartufai* (hunters) as watchmen of the territory, landscape, and history. Without them, apart from finding the truffles, many business activities such as festivals or touristic experiences would not be possible.

The purpose of this study is to report the opinions of the truffle hunters on what improvements can be made in the preservation of areas where truffles grow. It also showcases the traditional knowledge associated with conservation and collection of the white truffle. This work has been written based on the opinions of the *tartufai* mostly belonging to the main Italian associations such as the Federazione Nazionale Associazioni Tartufai Italiana (FNATI) and the Associazione Nazionale Città del Tartufo (ANCT), whom we thank for the help given. After an introduction on the truffle sector followed by a brief explanation of interview methods, the chapter will provide a summary of the responses of the participants, followed by a series of actions to be taken based on these answers. The areas where the interviews were conducted (Piedmont, Tuscany, and Emilia-Romagna) are among the most productive in Italy (Table 4.1).

1.2 Where Are the Truffles Found?

Compared to other edible truffles, the *Tuber magnatum* variety (Fig. 4.1) is usually found in Italy, especially northern and central areas (Fig. 4.2), France, southern Switzerland, and the Balkans (Hall et al. 1998).[1] They grow mostly in association with trees in humid areas near rivers or canals which are dug by farmers to irrigate or manage rainwater. According to a report on the truffle sector by Italy's Ministry of Agriculture, Food Sovereignty and Forests entitled, "Piano Nazionale Della Filiera del Tartufo (PNFT) 2017–2020" (MASAF 2018), ectomycorrhizal species that are hosts for *Tuber magnatum* include *Alnus cordata Desf* (Italian alder), *Corylus avellana L* (hazelnut), *Carpinus betulus L.* (hornbeam), *Populus alba L.* (white poplar), *P. tremula L.* (European trembling aspen), *P. nigra L.* (Lombardy poplar), *Quercus cerris L.* (turkey oak), *Q. ilex L.* (holm oak), *Q. pubescens Willd.* (downy oak), *Q. robur L.* (English or common oak), *Salix alba L.* (white willow), *S. caprea L.* (pussy willow), *Tilia cordata Mill.* (small-leaved lime), and *T. platyphyllos Scop.* (large-leaved lime). The sources interviewed in this study mentioned that the most productive hosts are *Quercus* spp., *Populus* spp., and *Salix* spp.

Table 4.1 Basic information of the study area

Country	Italy
Regions	Piemonte, Emilia-Romagna, Toscana
District	
Municipality	Firenze, Siena, Pisa, Grosseto, Lucca, Bologna, Parma, Piacenza, Sasso Marconi, Casale Monferrato, Alba, and others
Size of geographical area (hectare)	48,000 hectares
Dominant ethnicity(ies), if appropriate	Italians
Size of case study/project area (hectare)	
Dominant ethnicity in the project area	
Number of direct beneficiaries (people)	22,370
Number of indirect beneficiaries (people)	250,000
Geographic coordinates (latitude, longitude)	North limit 42°37'15.1"N 12°42'49.1"E 42°37'15.1"N 12°42'49.1"E South limit)

[1]According to Elena Salerni at the "Coltivare il Tuber Magnatum" conference on 26 November 2023 in San Miniato, some *Tuber magnatum* were also found in Thailand.

Fig. 4.1 *Tuber magnatum Pico* (white truffle), called *Tartufo Bianco Pregiato* or *Tarfufo Bianco d'Alba.* (Photo Credit: Associazione Nazionale Città del Tartufo)

1.3 How Are Truffles Found?

Truffles are hunted primarily by professional truffle hunters, known as *trifulau* in the Piedmont region of Northwest Italy (Fig. 4.2), but the general term in Italy is *tartufai* or *cavatori*. The practice of truffle hunting dates back to the ancient Roman civilization when people believed that the truffle was created when lightning struck the damp earth. Because of their scarcity and mystic nature, truffles have been a sought-after delicacy since their discovery. The *cavatori* have always used animals with an amplified sense of smell to aid in the forage: the most common are pigs and dogs. Pigs or wild boars are natural foragers and root for food in the ground with their snouts, but it was primarily the female pigs that were used to find truffles. This was because the natural sex hormones of the male pig were believed to be similar to the smell of the truffles, thus immediately attracting female pigs to the scent of truffles. Because truffles contain compounds that resemble testosterone found in pigs, some people think of truffles as an aphrodisiac. The pigs can detect truffles by their aroma alone, even if growing deep underground. The problem was that, with their long snouts, pigs damaged many of the truffles they dug up, if not consuming them completely. They also dug larger holes than necessary, damaging the truffles and tree roots.

The Italian government prohibited the use of pigs in truffle hunting in 1985 as the damage caused had the potential to reduce the production rate over time. In the modern era, dogs are now the *cavatori*'s primary companions (Fig. 4.3). Dogs undergo a long training process before they are ready to start hunting. Truffle hunting is highly regulated to ensure the truffles' growth and reproduction. Hunting

Fig. 4.2 Map of Italy with the areas where most white truffles are found in yellow. Upper-left red circle shows Piedmont and centre red circle marks the Tuscany and Emilia-Romagna regions where interviewed hunters operate. (Source: from interviews and data in Hall et al. (1998))

Fig. 4.3 Truffle hunting dogs. (Photo Credit: Associazione Nazionale Città del Tartufo)

hours and seasons are set by the Ministry of Agriculture, Food Sovereignty and Forests, but regional governments can alter some of the regulations. All *cavatori* need to pass an exam and obtain a licence. Most belong to a local association. In addition to professional hunters, truffles are sometimes harvested by hobbyists or amateur collectors who have licences but hunt only for personal consumption or to sell to local markets.

1.4 Who Owns the Wild Truffle?

Non-cultivated truffles grow only when special conditions are met. The areas where they grow are called *tartufaie* in Italian. Italian law (Act No. 752 of 16 December 1985) clearly states that the truffles belong to the owners of the lands where the *tartufaie* are located. Because truffle hunters can traditionally forage on any property that is not fenced, *tartufaie* owners who wish to harvest themselves have to seek an authorization and put up a sign to warn the hunters not to take any truffles. When this happens, the area is then closed to unauthorized people. Local councils monitor and give permission for the creation of closed *tartufaie*. As this is happening more frequently than before, foraging associations are concerned that areas for free searching will become increasingly rare for members.

When foraging season is open in non-cultivated and non-closed areas or public woods, truffles can be taken by individuals who have passed the exam and have a licence. *Cavatori* need to respect several rules such as not using pigs and filling up the holes in the ground made by their tools. Many *cavatori* say that some new people who lack the necessary knowledge are leaving large holes open or picking truffles at the wrong time, thereby failing to ensure reproduction. Although there is no system in place to reward the owners of *tartufaie*, such as a payment for ecosystem services (PES) mechanism, there can be private agreements (see below regarding differing interests of stakeholders in Sect. 2.3).

1.5 The Truffle Sector

It is difficult to determine the exact worth of the white truffle (*Tuber magnatum*) sector as it is subject to market fluctuations and can vary greatly depending on the availability and quality of the truffles. However, white truffles are one of the most valuable and highly prized culinary ingredients in the world, with prices ranging from hundreds to thousands of euros per kilogram. The luxury nature of white truffles and their scarcity contribute to their high value, making the white truffle sector a lucrative market for producers, traders, and restaurateurs. *Tuber magnatum* is extremely difficult to farm and can still mainly be found in the wild. Besides the truffles' low yield, they are seasonal and have a short shelf life, and quantities harvested have been in decline for decades. Farmed truffles or non-scented truffles are found more easily, and their cost is much lower.

When a white truffle is whole, it has the most value. In 2022, prices ranged between 5000 and 6000 EUR per kg. Small truffles or fragments were sold in 2022 at prices between 3000 and 4000 EUR per kg. Less precious truffles ranged in cost from 100 to 300 EUR per kg in 2022.

The *cavatori* can sell truffles to consumers, restaurants, or processors. According to several sources interviewed, there were about 200,000 truffle foragers in Italy as of January 2023, although data from the Ministry of Agriculture in 2016 (MASAF 2018) states this figure at just over 80,000 registered *cavatori* in Italy.[2] It is very difficult to track down *cavatori*'s revenues as most of them do not have companies and as self-employed individuals do not necessarily reveal their income related to truffles.

The majority of the *cavatori* are concentrated in the north (Piedmont, Lombardy, and Emilia-Romagna) and centre (Tuscany, Umbria, Marche, and Abruzzo) of Italy. This research focuses on those in Piedmont, Tuscany, and Emilia-Romagna. The truffle industry employs more people than the foragers themselves. According to the truffle market outlook for 2023–2033 (Fact.MR 2022), demand for truffles worldwide will grow within the next decade to reach 800 million USD from today's 340 million USD. Along with the demand increase, 60% of this trade will be represented by the black truffle. If this is the case, more truffles will be needed, most likely cultivated ones as the white truffle only grows in the wild. Besides restaurants and farmers, truffle tourism is increasing dramatically and generates revenues in the millions of euros, much more than fresh truffle sales.

1.6 The Impact of Truffle Hunting on the Local and Global Economy

The impact of the truffle industry on the local economy is very positive but difficult to assess in terms of numbers. Marone (2011) (Fig. 4.4), shows that in Tuscany most fresh truffles are acquired by restaurants (39%) or the processing industry (39%), who will add value to the product. In terms of geographical distribution, 32% of sales in Tuscany take place outside Italy, 39% in Italy, and 29% in the Tuscan region (Marone 2011). These are revenues directly and only related to truffles and truffle processing.

[2] The ministry issues permission to hunt, and thus we know that at least 80,000 were granted a permit according to data released in 2016. This is incomplete data, as not all hunters apply for a permit. Data from truffle associations suggest that the number of hunters could be as high as 200,000, but only 70,000 would be active in Italy. At least 7000 have paid taxes for selling wild truffles. Non-professional foragers are taxed 100 EUR per year if their revenues do not exceed a certain amount.

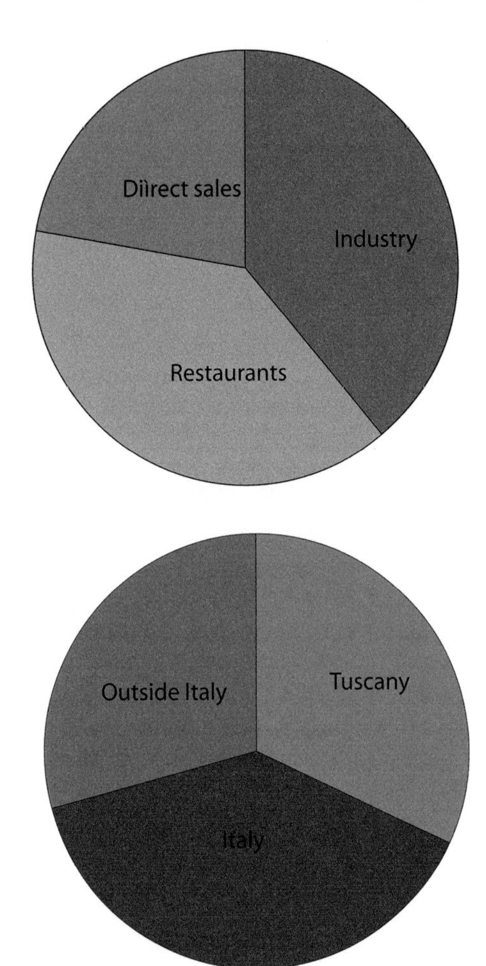

Fig. 4.4 A study in the Tuscan region by Enrico Marone (2011) shows truffle revenues: the top pie chart is by industry. When a truffle is sold by a hunter, 39% goes to companies who process them and 39% to restaurants. The rest (22%) is sold directly to end users. Of these, most end up in Italy, but many go abroad (32%), as seen in the pie chart below. Tuscany and Piedmont, as two of the main tourist destinations, also sell truffles imported from other Italian regions (and also from other countries, such as the Balkan area) given the insufficient local supplies

The tourist sector is much harder to assess. Tourists come to Italy, and especially to the truffle regions, because of the food. So, truffles, wine, cheese, and meats are part of Italy's attraction. When truffle hunting season is in full swing, famous fairs and festivals dedicated to the unique flavour of the *tartufo bianco*, or white truffle, are held everywhere from Tuscany to Piedmont. Across Italy, there are more than 115 regional events dedicated to truffles, mainly in Piedmont, Molise, Tuscany, Marche, Umbria, and Emilia-Romagna. The International Alba White Truffle Fair and the National White Truffle Exhibition in San Miniato are very important events and attract many local and international visitors (Fig. 4.5). The truffle festivals are a gourmet, gastronomical festival to honour the prized white truffle. They highlight traditional recipes, local businesses, and authentic products that boost the local economy and give attendees an authentic truffle experience.

People from all around the world travel to the Tuscan or Piedmont countryside to immerse themselves in the medley of flavours, smells, and colours that come

Fig. 4.5 Local truffle event in Piedmont. (Photo Credit: Associazione Nazionale Città del Tartufo)

together around the celebrated truffles. Festivals work to create and pass on cultural identity. They can help bind people to their communities, foster and reinforce group identity, and keep cultural heritage alive (Figs. 4.5 and 4.6). Today, one cannot walk the streets of Florence or Alba without passing a truffle shop, a restaurant with a truffle menu, or a tourist centre promoting truffle hunting. Tourists enjoy trying different truffle oils, pasta varieties, and pairings at local restaurants, festivals, and shops. Truffle hunting experiences are often the primary motive for tourists to come to Tuscany or Piedmont. Truffles are a perfect synthesis of the identity, community, and gastronomy of many Italian regions. All tourism businesses benefit from the truffle presence, with the white being the star of truffles. Therefore, the reduction in white truffle production is a threat to the tourism sector, especially in the regions and areas where truffles are the main attraction. There would be a significant economic impact specifically on the hotel and restaurant business.

One third of hunted truffles are exported, not to mention the processed products that are mainly exported and generate even more revenue and business worldwide. Main export markets are North America and Northern Europe as well as Asian countries such as Japan, China, South Korea, Vietnam, and Singapore. Some truffles are also cultivated in Asia by a new growing industry.

Fig. 4.6 People eating at a truffle event. (Photo Credit: Associazione Nazionale Città del Tartufo)

2 Interviews with Truffle Hunters

To assess the factors hindering truffle production and foraging, 14 interviews were conducted among truffle hunters and several presidents of truffle associations. Figure 4.7 shows the number of members of the different associations. Most of them are small with <1000 members. Truffle hunters join associations for a variety of reasons, such as to be in a local community, to get help with paperwork, to have access to private *tartufaie*, and to help maintain the ecosystems. Many hunters do not join any associations and therefore their activity cannot be tracked.

Overall, we were able to gather data connected to more than 22,000 *cavatori* via their representatives. We asked the representatives of the *cavatori* whether truffle production was stable, increasing, or declining for each of the past 5 years. Then we asked them to list the reasons for any decline in order of importance. Finally, we asked the interviewees to suggest actions to be taken to reduce that decline.

2.1 Decline of Tuber Magnatum Harvest

As most sources reported a decline (see Table 4.2), the white truffle (*Tuber magnatum Pico*) harvest was shown to be declining yearly in the majority of areas we explored. The areas that have experienced the most decline are the completely wild ones. The private or semi-private areas that have been taken care of either by

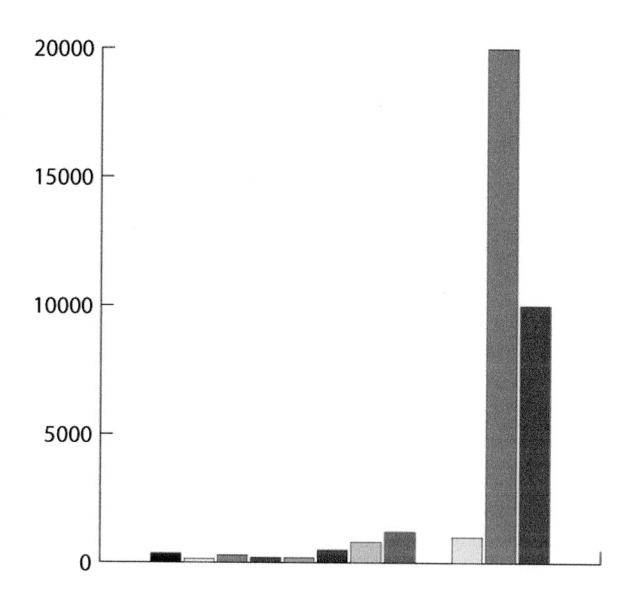

Fig. 4.7 Number of members per association interviewed. The total number of members is 22,373 registered truffle hunters. Two sources have overlapping members so the graph does not add up to 22,373

Table 4.2 The 12 interviewees in three regions were asked whether the truffle harvest was stable (→), declining (↓), or growing (↑) year on year for the past 5 years. Most saw a decline. Of the 14 interviews, two sources that overlapped with others were not included

Interviewees (12) by region	2022	2021	2020	2019	2018
Tuscany	↓	↓	↓		↑
Tuscany	↓	↓	↓	↓	↓
Tuscany	↓	↓	↓	↓	↑
Tuscany	↓	↓	↓	↓	↓
Tuscany	↓	↓	↓	↓	↓
Tuscany	↓	↓	↓	→	↑
Emilia-Romagna	↑	→	↓	↓	↑
Emilia-Romagna	↓	↓	↓	↓	↑
Tuscany	↓	↓	↓		↓
Emilia-Romagna	↓	→	↓	↓	↓
Emilia-Romagna	↓	↓	↓	↓	→
Piedmont	↓	↓	↓	↓	↓

farmers or the associations have experienced less decline. Compared to *Tuber magnatum*, other types, most of which are cultivated truffles, did not see such a sharp decline and the volumes of their harvest were higher.[3]

2.2 Main Reason for Harvest Decline: Lack of Water and Torrential Rains

All 12 sources said that the most important factor in truffle reproduction and growth is rain. Several said there are not enough proper rains (long, but not too violent). Rain in the spring and occasionally in summer is necessary for the truffles. Interviewees explained that incidences of no rain or heavy torrential rains have been more frequent lately. Given the fact that rainfall cannot be controlled, it is extremely important that water management is done properly. Until 1950, many farmers were sharecroppers in central Italy. Their livelihoods were dependent on the land, so they took care of everything in all their fields. Landlords also asked sharecroppers to maintain canals and waterways as well as maintain woodlands by keeping them clean. Today's farming is mainly concentrated on cost cutting and economies of scale. Water management is expensive and a long-term endeavour, and therefore not seen as a priority. The combination of either no rain or torrential rain, alongside either obstructed or non-existent waterways and canals, is a catastrophe for truffles as areas are wiped out or buried with debris and landslides, hindering truffle growth.

2.3 Other Reasons for Harvest Decline

The interview results showed the following factors as other reasons for the decline in harvest of the white truffle:

2.3.1 Loss of Biodiversity and Habitat

Eleven of 12 sources said that declining biodiversity is a problem for a variety of reasons. Most said that the trees where the truffles grow had been cut or not maintained, and two reported that invasive tree species such as *Robinia pseudoacacia* (black locust or false acacia) and *Ailanthus* are replacing the host trees and therefore reducing the truffle habitat. Three noted that introduced non-native species of truffles could take over the precious white truffle, which spreads with more difficulty.

[3] According to Reyna and Garcia-Barreda (2014), on the one hand the annual European production of *Tuber melanosporum* averaged around 58 tons per year for the period 2003–2012. On the other hand, the Italian production of *Tuber magnatum* in 2005 was one ton according to Marone (2011).

The introduction of different and foreign truffles can also spread pests and diseases, such as fungal infections.

The reasons given for the loss of trees are multiple. These range from human activities such as eliminating woods in between parcels or not engaging in good practices for tree conservation (e.g. pruning trees regularly, cutting the underbrush when it gets too big, and digging canals to eliminate excess water), to wild animals who damage trees, climate change, and lack of water.

2.3.2 Tree Management

Tree management was quoted as one of the top reasons for the reduction in the white truffle harvest. Truffles grow in two main areas: wild woodlands and moist areas next to cultivated fields. Wild woodlands used to be maintained for firewood and were cleared frequently leaving some of the trees for reproduction. Today firewood is cut without the truffle habitat in mind, and too few trees are left, therefore decreasing shade and leaving the soil too dry and hot for white truffles. At the same time woodcutters do not pay attention to soil health as they use large machines to remove the wood, damaging the soil and sometimes removing part of it when passing with large trucks. This activity was much less invasive in the past as it was usually done manually.

In the areas next to cultivated fields and near rivers, tree management is either absent or done in an inadequate way. Some of the arable fields are cultivated to the maximum extent with deep ploughs so that the place for trees around the fields is either reduced or eliminated. In other areas, trees are not maintained and especially poplars grow old and die and so do truffles. In areas near riverbeds, tree maintenance is usually not kept up as farmers need to cut costs in a very competitive market. In some of the truffle growing areas, for example in the province of Florence in Tuscany, there is a property tax (paid by every property owner) that goes to maintaining waterways and river areas. This tax is collected by the region and given to what are called *consorzi di bonifica* (land reclamation consortia). Most of the time, their activities are not performed with respect for the truffle habitat as there is a lack of TEK concerning truffles. It is cheaper to use large machines to clear areas than to manually remove dead trees and debris, while paying attention not to damage the soil. Large machines can destroy the ecosystem that has nurtured trees with truffles for decades if operators do not respect the trees that potentially have truffles in their roots. There is a need for communication between the consortia and the truffle associations where the latter is able to guide and control consortia activity in truffle growing areas. The mapping of the areas that is being done by the region will help when it is finished.

2.3.3 Agricultural Activities

Agricultural activities can be harmful in several ways to truffle ecosystems. Use of herbicides, pesticides, and especially fungicides close to riverbeds and woods hinders truffle growth. Tilling the soil too deeply along the rivers and canals also damages roots where truffles grow. Leaving underbrush growing, not maintaining the canals in cultivated fields, and allowing dead trees or dirt to accumulate on the sides of cultivated fields asphyxiates tree roots and truffles. Furthermore, the size of cultivated fields has increased in the past 70 years. Fields were much smaller in the past leaving more areas around each field in which trees could grow. Now, because of economies of scale and cost cutting concerns, fields are bigger, small parcels are consolidated, and the trees around them are either cut or not maintained.

2.3.4 Wild Animals

A few sources reported damages to truffles by wild boars. Truffles located in wild areas suffer more than those in areas sheltered from wild animals. The number of wild boars has increased in the past decades resulting in damage to all sorts of agricultural practices such as grape and wheat cultivation. A few sources showed a mixed view about wild boars. A couple of interviewees said wild boars are useful when they dig the soil: if they do it lightly, they contribute to spreading the truffles. Four others noted only negative impacts as the boars are too many. The Tuscan region has extended the wild boar hunting season to all year to help reduce the number. However, not all stakeholders agree with this measure. For example, hunters' associations prefer to have plenty of boars to hunt and do not necessarily want to reduce their numbers too much. Animal welfare organizations protest if hunting legislation is too permissive. Local politicians try to please the different groups, and strong actions are not taken.

2.3.5 Loss of TEK

In general, the loss of TEK is an ongoing problem in foraging. The main issues are the timing of foraging and the way in which truffles are removed from the ground. *Cavatori* who do not know how to collect truffles might remove them too early or they might over-harvest, hindering reproduction. Taking them before they develop their typical smell decreases the value and quality of the product. The other issue that sources reported is that not all foragers cover the holes left by their tools, leaving roots exposed or the soil open. In that case, again, reproduction is hindered.

In terms of the trees, sources reported that old farmers know how to keep trees healthy and renew them so that they can always host truffles, while those on large farms now do not even think about this issue.

2.3.6 Differing Interests of Stakeholders

In the end, a very large and unresolved problem is the fact that the interests of different stakeholders do not coincide. Truffle hunters want to find truffles, but they do it mostly on other people's land without giving them anything. Farmers want to cultivate with the lowest possible costs and do not particularly want to do any favours for truffle hunters as they gain nothing from it. Maintenance and water management companies need to save money in their operations, and caring for truffles increases costs. The tourist sector wants a steady flow of truffles at a reasonable price. Local politicians would like to please everyone but often do not have the resources or solutions. So far, the best working options to solve this problem have been either a contract between the landowner and local associations or individual hunters where the landowner gets some benefit, or the privatization of truffle areas with fences.

3 Possible Solutions

3.1 Proposed Actions To Be Taken

As we have seen, the habitat of the white truffle is endangered. As benefits from the truffle harvest usually do not go to the owners of the fields, it is particularly difficult to find an agreement between beneficiaries (including associations and individual hunters) and landowners. Sources were asked what would be needed to help recover the truffle ecosystem and who should do it. Rain at the right moment was a common answer but not achievable by intention. In terms of human interventions, suggested actions to improve truffle growth according to the associations included planting appropriate trees, managing water canals and river areas, cutting old trees or dead trees and replacing them with new ones of the same varieties, reducing the number of wild boars, clearing areas near rivers without damaging the soil, avoiding debris deposits, and planting trees with roots already with mycorrhiza.

3.2 Ongoing Actions

3.2.1 Actions by Associations

Most of the associations had already started to act using the fees members pay to register. They are the main actors and the main intermediaries between all stakeholders. If their actions fail, the only remaining actors are private enterprises that are not the keepers of TEK. The associations have rented land or made agreements with landowners to manage their truffle areas by keeping them clean of accumulated dirt and debris, clearing canals, planting trees, or cutting them to renew them, as well as

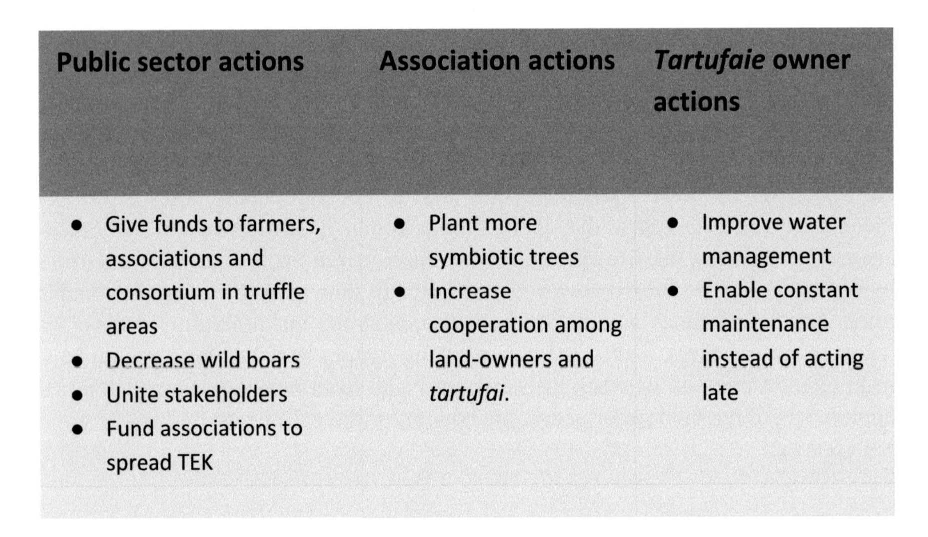

Fig. 4.8 The role of the associations as the keepers of TEK: here is what they suggested as actions to be taken

closing off some areas with fences to keep animals out. Interviews showed that *tartufaie* that are managed well produce more (Fig. 4.8).

One of the interviewees who has a farm with *tartufaie* has fenced the areas and managed the trees and therefore has better yield. Some other associations have created private *tartufaie* as well due to the fact that nobody wants to pay for maintenance unless they are sure to reap the benefits. If this becomes the general trend, public *tartufaie* would disappear in favour of private ones. The whole sector of free truffle searching would slowly disappear, and TEK would only be kept by the owners of the *tartufaie*. Many of the sources were against transforming truffle searching into a private business. At the same time, they said that members of the associations do not want to pay for the maintenance of the fields unless they have priority access to the *tartufaie*: "*Why would I do all that work to have somebody else get the truffles?*" While it is clear that managed *tartufaie* are less at risk of harvest reduction compared to natural ones, it is hard to determine who should do the work of maintaining them. If production were to come only from private ones, the traditional truffle hunting knowledge would be at risk along with many of the social and tourism activities surrounding it, so that should not be the preferred solution (Fig. 4.9).

Fig. 4.6 Another step taken by truffle associations at the national level has been a large communication campaign, mainly concerning the traditional knowledge, that should help discussions between landowners and truffle hunters. By raising interest and working on public opinion, the associations hope to make landowners feel more responsibility towards the community. If associated with some sort of payment for ecosystem services (PES), this action could improve truffle harvest and therefore increase business output.

Agricultural company actions	Association/ hunter actions	Industry actions
• Fenced areas • Plant trees • Improve water management • Constant maintenance	• Keep areas clean • Plant trees • Increase cooperation among landowners and *cavatori*.	• Diversify sourcing • Increased usage of synthetic molecules instead of real truffles
Expected results: • Private tours • Product sales • Access restricted	**Expected results:** • No change in financial benefits • Maintaining tradition	**Expected results:** • Increased sales • Lower prices • Lower quality

Fig. 4.9 Actions to be taken, list of most common answers (mentioned by more than five interviewees)

3.2.2 Actions by the Public Sector

The associations pointed out that public sector actions have mainly been undertaken by local governments. In fact, the public sector is probably the only stakeholder that would be able to take care of the *tartufaie* for the sake of all stakeholders. In the PNFT 2017–2020 report (MASAF 2018), the problems of truffle ecosystem degradation reported by the interviewed hunters are clearly stated, and several solutions are analyzed, such as creating an environmental fund to invest in *tartufaie* preservation. However, the report proposes that the national government delegates the power to regional governments to find and develop a strategy for the preservation of the truffle ecosystem.

In fact, sources reported on local main action points and associated challenges. For instance, a truffle hunter from Piedmont said that the regional government started to pay farmers to maintain trees, but since farmers were not checked, they took the money but did not maintain the trees most of the time. Sources from Tuscany said that they tried to work with the consortium in charge of river areas to make positive impacts on the truffle ecosystem. However, only in very limited areas did the owners of the fields, the truffles associations, and the workers of the consortiums cooperate, so coordination among stakeholders is needed.

Another action point is cataloguing the areas more precisely. Local governments are mapping the *tartufaie* so that at least the areas are known. This represents a good start, and will in the least reveal areas for interventions.

Despite some local actions, when talking to truffle hunter associations, it does not seem that there have been any changes in the field and government programmes are still largely theoretical. Cooperation between the owners of the land and the *cavatori* needs to be increased so that the owners get some benefits that in turn push them to maintain the fields. Subsidies would help and can only work if reinforced, as farmers or maintenance companies do not benefit from taking care of the free *tartufaie* as the sales of truffles found by external truffle hunters do not benefit them in any way. The associations also have problems with managing *tartufaie* in public woods. Unless they have public funding, members of associations would only pay to manage private *tartufaie* to grant themselves the right to forage them. A possible solution would be for the local government to pay and instruct semi-public structures to take care of the truffle areas for the benefit of the community and distribute the benefits to the owners of the land, who in theory are also the owners of the truffles.

4 Conclusions

From all the interviews, managing the *tartufaie* and therefore continuing to have a steady flow of different truffles (not only the cultivated ones) increases truffle yield. That, in turn increases business benefits for the *cavatori* and all the businesses after them (restaurants, tourism, and industry). As demand is growing from the industry and tourism, truffle hunters can sell more if there is more to sell. In this case, negotiations among stakeholders are a key issue. When a landowner and a group of *cavatori* agree and maintain an area by sharing the revenues of truffle sales, they all gain in terms of revenue. The *cavatori* knowledge should be shared with farmers so that they do what is needed to maintain the truffle areas, but farmers need to be rewarded for their effort. Local governments should help with truffle friendly legislation. The industry is selling a beautiful story involving humans, dogs, and scented truffles, recognized by UNESCO. If the areas continue to supply the real products, they can also gain in terms of business and at the same time contribute to maintaining truffle (and tree) biodiversity.

Acknowledgements I would like to thank the following associations for their participation in this work: Laura Giannetti Truffle expert and Consigliere Ass. Tartufai Colline Bassa Val d'Elsa, Fabio Cerretano, presidente F.N.A.T.I. - Federazione Nazionale Associazioni Tartufai Italiani, Rebecca Narcisa Gualandi and Debra Williams-Gualandi for general help, Renato Battini and Guido Franchi from Associazione Tartufai Delle Colline Samminiatesi, Giovanni Gallerini from Ass. Tartufai delle colline della Bassa Val D'Elsa, Roberto Vanni from Ass. Tartufai Senesi, Roberto Papalini from Associazione Tartufai Dell'Amiata – Castell'Azzara, Daniele Montigiani from Tartufai del Garbo, Luigi Soprani from Associazione Tartufai Sasso Marconi, Lino Costi from Associazione Tartufai Reggiani, Guido Manfredi Rasponi from Az. Agricola Barbialla Nuova, Walter Pieroni from Associazione Parmense Raccoglitori Tartufi, Giuseppe Crescente from Associazione Nazionale Tartufai Italiani Regione Emilia Romagna, Antonella Scaglia from Associazione Tartufai del Monferrato Piemonte, and Antonella Brancadoro, Direttore Associazione Città del Tartufo.

References

Fact.MR (2022) Truffle market outlook (2023–2033), viewed 26 January 2023. Retrieved from https://www.factmr.com/report/truffles-market

Hall IR, Zambonelli A, Primavera F (1998) Ectomycorrhizal Fungi with Edible Fruiting Bodies 3. Tuber magnatum, Tuberaceae. Economic Botany 52(2):192–200

Marone E (2011) *La filiera del tartufo e la sua valorizzazione in Toscana e Abruzzo*, Firenze University Press.

MASAF (2018) (PNFT 2017–2020), *Piano nazionale della filiera del tartufo 2017–2020*, Ministero delle politiche agricole alimentari e forestali, Rome, Italy, viewed January 26 2023, Retrieved from https://www.politicheagricole.it/flex/cm/pages/ServeAttachment.php/L/IT/D/d%252Ff%252F2%252FD.ae1e3c2ae5bff1e25deb/P/BLOB%3AID%3D11100/E/pdf

Reyna S, Garcia-Barreda S (2014) Black truffle cultivation: A global reality. Forest Systems 23(2):317–328

Guido Gualandi Teaches History of Food in the Mediterranean at Gonzaga University in Florence and Accent International. He has worked as an archaeologist in France, Italy, and the Middle East. He owns a farm in Tuscany and is a member of the Ancient Grains Community in Montespertoli, Italy.

Chapter 5
Nature-Positive Approaches to Sustainable Tourism Development in the Panchase Region of Nepal

Dambar Pun, Aashish Tiwari, Rebecca Gurung, Hum Bahadur Gurung, and Samuel Pun

Abstract Mass tourism has been recognized as a key factor in undesirable environmental and social impacts on host realms that work against sustainable development and involve negative impacts like environmental and social degradation, economic inflation and leakage, loss of habitats and wildlife species, and various social problems. The Panchase Region of Nepal needs to be developed as a model ecotourism destination with minimal impacts on ecosystem services, biodiversity loss, pollution, and climate change.

'Nature-positive' approaches not only halt the loss of ecosystems and the services they provide, but also assist in recovery and replenishment of ecological systems through participation and collaboration among different stakeholders at multiple dimensions leading up to end uses. This chapter will explore a sustainable tourism development model for the operation of an ecosystem restoration-oriented private tourism business in the Panchase region. Key informant interviews, consultations, and focus group discussions on the nature-positive approach were conducted. The pristine landscape of the Panchase mountain region supplemented by the World Peace Biodiversity Park has attracted more visitors for trekking and homestays, which has placed tourism development on the verge of exploiting the natural environment and resources.

Good governance practices can be replicated for private sector tourism with socially and ecologically balanced sustainable tourism approaches that are nature-positive and embody principles of ecotourism and socio-ecological production

D. Pun (✉)
Back to Nature, Pokhara, Gandaki, Nepal

A. Tiwari
Conservation Development Foundation, Pokhara, Gandaki, Nepal

R. Gurung
Himalayan Sustainable Future Foundation, Kathmandu, Bagmati, Nepal

H. B. Gurung
BirdLife International Asia, Tanglin International Centre, Singapore

S. Pun
Backyard by Back to Nature, Kathmandu, Bagmati, Nepal

M. Nishi et al. (eds.), *Business and Biodiversity*, Satoyama Initiative Thematic Review, https://doi.org/10.1007/978-981-97-7574-3_5

landscapes (SEPLs). Integration of these approaches will create a positive environment for sustainable tourism development and tourism entrepreneurs such as Back to Nature, leading to the utmost preservation and protection of critical habitats in the Panchase Region of Nepal.

Keywords Biodiversity conservation · Back to nature · Ecotourism · Mass tourism · Nature-positive · Restoration · Sustainable development

1 Introduction

Business activities have impacts on multiple dimensions of nature and social systems, from the climate and biodiversity, to communities in production landscapes. Organizations often address these dimensions independently (e.g. climate and biodiversity strategies are typically stand-alone), but this brings the risk of actions across different dimensions undermining rather than reinforcing each other, such as decarbonization strategies having the potential to inflict serious harm on biodiversity (Pörtner et al. 2021). Plantation-based carbon offsets, increased biofuel consumption, and production of the raw materials needed for a renewable energy transition could all conflict with biodiversity goals (Sonter et al. 2020; Sen and Dabi 2021). Delivering a nature-positive strategy involves contributing to environmental improvement across the dimensions relevant to a business (e.g. biodiversity, carbon, and water) using an integrated approach that addresses impacts coherently without exacerbating other environmental or social risks and manifests in businesses investing in solutions that contribute to multiple goals simultaneously (e.g. green infrastructure, nature-based solutions focusing on the restoration of natural ecosystems, or combined target-setting) (Bull et al. 2020; WBCSD 2021a). The heads of 14 conservation organizations described "nature-positive" as halting and reversing nature loss by 2030, measured from the baseline of 2020 (Locke et al. 2021). The Global Goal for Nature Group developed an umbrella goal for the post-2020 Global Biodiversity Framework (GBF), which stated that nature-positive goals cover actions to preserve networks of protected areas, different species of wildlife and plants, and agriculture as a whole-of-society approach. Business plays an important role in the post-2020 GBF by conducting activities whose benefits outweigh their negative impacts, from restoration to benefit sharing with Indigenous people (WBCSD 2021a, b).

1.1 Tourism in Nepal

In Nepal, tourism is one of the businesses that depends on nature. The UN World Tourism Organization accounts a rise in tourists to South Asia, from 2.9 million in 2021 during COVID-19 to 34.9 million in 2022 after the global pandemic (UNWTO

2023). Prior to COVID times, Nepal saw over one million tourist arrivals in 2019. During COVID and given all the restrictions, just 35,893 tourists visited in 2020 (MoCTCA 2020). Currently, the trend shows an increase of 398.8% from 15,549 tourists in 2021 to 61,701 in 2022 (MoCTCA 2022). Scenic countries like Nepal with its epic mountains have experienced fast economic growth due to tourism, but it is a double-edged sword: tourism brings economic prosperity but also challenges for managing protected natural areas (Simmons 2013; Davlatbek and Doniyorbek 2023). Further, tourism in developing countries can impact conservation on private land too (Buckley 2011).

There are various types of tourism, and among these ecotourism has been understood as an opportunity to improve local livelihoods and conserve nature (Samal and Dash 2023). Ecotourism is also considered to be a tool to achieve sustainable tourism (Yogi 2010). It has been defined by a framework that includes "nature-based products and markets, sustainable management to minimize impacts, financial support for conservation, and environmental attitudes and education of individuals" (Buckley 1994, p. 4). When nature-based tourism helps to minimize the negative impacts of tourism, makes it more sustainable, and supports conservation and changes in the economic behaviour and education of individuals, then it can be called ecotourism (Buckley 1994).

1.2 Development and Trend of Ecotourism in Nepal

After the 1950s, tourism was developed in Pokhara and other natural areas of Nepal, then taking on an economistic orientation. Development was envisaged as mere economic development in contrast to the global movement for sustainable development (Regmi and Walter 2017). After the destruction of ecosystems, wildlife habitats, cultural degradation, displacement of Indigenous people, and decrease in quality of services were brought by conventional mass tourism in many countries, alternative types of tourism like nature-based tourism, responsible tourism, adventure tourism, green tourism, village tourism, wildlife tourism, culture tourism, educational tourism, agrotourism, and scientific tourism were sought after by the tourism industry (Yogi 2010). Ecotourism is also a type of alternative tourism which has the potential to uplift local livelihoods and conserve nature (Samal and Dash 2023). The first ecotourism project in Nepal was the Ghalekharka-Sikles Eco-Trek area launched in 1992 (Nyaupane and Thapa 2004). A study on the impacts of tourism in the internationally renowned Annapurna Conservation Area showed fewer negative and more positive impacts on economy and environment (Nyaupane and Thapa 2004). Ecotourism is environment friendly and if implemented properly also helps achieve the sustainable development goals (Ojha 2020). Ecotourism helps to increase job security and promotes the use of local products, conservation of local biodiversity, and Indigenous cultures (Anup 2017).

Biodiversity is under threat from environmental factors (e.g. climate change and natural disasters) and anthropogenic factors (such as infrastructure and increased tourism development). Private sector tourism involvement in restoration activities, implying the involvement of communities in ecosystem services trade-offs through implementation of nature-positive approaches with non-consumptive uses of nature, can create a suitable enabling environment to develop ecotourism in the Panchase landscapes. Panchase is home to a wide range of traditional communities (including forest dwellers, fisherman, and farmers—Indigenous women and locals). These communities have a rich heritage and knowledge of more traditional and sustainable lifestyles, agricultural management, and farming practices. While the elder generation traditionally has developed a more sensitive, holistic approach to the biological diversity of their environment, the younger generation tends to focus on the income generation of tourist infrastructure (hotels, restaurants, etc.) (Pun et al. 2023). This new paradigm puts nature and critical habitats on the verge of degradation from human-related tourism interventions and activities. The Panchase region provides ecosystem services like timber, medicine, sand gravel, and fibres with provision and support for communities and private tourism sectors.

2 Panchase Protected Forest and Panchase Region

Pokhara city, known as the tourism capital of Nepal, lies in the Kaski district and is renowned as a lake city. Phewa Lake, which was designated as a Ramsar Site (as part of the Lake Cluster of Pokhara Valley), is a wetland area of international significance. The main feeder and river of the lake, the Harapan watershed, originates in the Panchase Protected Forest (PPF) (Adhikari et al. 2018b). The PPF is a biodiversity hotspot in the mid hill region of Nepal, where forest species are found from 800 to 2517 m above sea level (Adhikari et al. 2018b; Paudel 2017). The forest comprises a high diversity of flora with 631 species of flowering plants dominated by orchids, which could be developed as an orchid sanctuary. In addition, two more genera of flowering plants were added as new flora recorded in Nepal (Bhandari et al. 2018). It is also home to 260 species of birds and 24 species of mammals, including two endangered species, one vulnerable, ten near-threatened species (Adhikari et al. 2018b) as well as eight species of bats (Bhandari 2020). It is situated at the junction of three districts—Parbat, Syangja, and Kaski, which is a connecting forest landscape between the lowland Chitwan landscape and the high Annapurna Conservation Area in Gandaki Province (Bhandari 2020; Paudel 2017). The Panchase forest and its peaks have religious significance for Hindus and Buddhists alike. The terrestrial ecosystem of the PPF and its slopes in all directions support different land use types, such as forest grazing and agricultural land. The steep hills and valleys are the major human residential spaces (Adhikari et al. 2018b; Paudel 2017). The network of forests in the PPF forms a complex system covering the larger landscape known as the Panchase Mountain Ecological Region (PMER), or simply the Panchase Region (PR). It provides ecosystem services like water, forest products, sand gravel, and fibres, provisioning and supporting communities with as many households as 14,807 (Adhikari et al. 2018a). The entire PR is subdivided into core and buffer regions, where the Government of Nepal has officially designated and outlined the boundaries of the core region (Fig. 5.1). However, the delineation of the buffer region boundary remains an ongoing challenge. This buffer region encompasses various stakeholders characterized by different altitudes and values, contributing to in situ conservation within the production landscape system, and involving activities such as agriculture and forestry, exemplified by the World Peace Biodiversity Park (WPBP) and Back to Nature (BtN) (Fig. 5.1).

The annual total ecosystem services (timber, water, and fuelwood) received in PR were estimated to be USD 685,212 in economic value for consumptive use, and USD 378,395 for non-consumptive use (of which, USD 234,395 was for religious activities and USD 144,000 for recreation) (Bhandari 2020). The PR provides recreation and tourism opportunities for 19,005 domestic, and 6335 international tourists and 25,340 pilgrims every year through hiking and walking in wooded trekking routes (Bhandari et al. 2018). These tourists can adversely affect the surrounding critical regions of the biodiversity hotspot and environmentally sensitive areas from unsustainable tourism development. Thus, these tourism activities should be ecologically, socially, and economically viable for the conservation and protection of biodiversity and susceptible landforms. Nature-positive approaches by the private

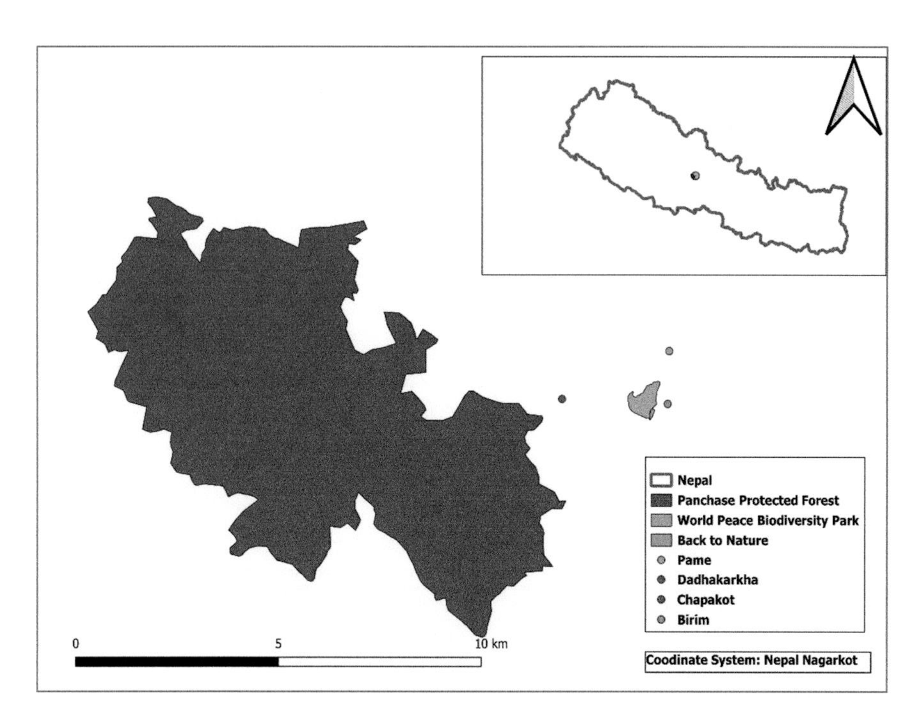

Fig. 5.1 Location of Panchase Protected Forest, World Peace Biodiversity Park, and Back to Nature in Nepal. (Source: Aashish Tiwari and Pun et al. 2023)

sector for sustainable tourism will be effective to conserve and protect the critical habitats of biodiversity that have been exploited by mass tourism. The socio-ecological production landscapes (SEPLs) in PR have broadened with the initiation of activities such as development of the World Peace Biodiversity Park, Pokhara, based on a participatory and collaborative approach in cooperation with the private sector, government, and communities, which continues to explore nature-positive approaches. In this chapter, we discuss the nature-positive ecotourism model of Back to Nature (BtN), a privately owned business that aims to sustain tourism and conserve biodiversity. This model could be replicated elsewhere in the PR (Fig. 5.1).

3 Back to Nature as an Ecotourism Enterprise

Back to Nature (BtN) is a privately owned ecotourism business venture, located on the banks of Phewa Lake. It is a traditional village-styled hotel with vernacular houses in a rural setting of about two hectares situated at 1019 m above sea level in a village called Dandakharka of Pokhara Metropolitan City within the PR (Table 5.1). BtN was established because the founder wanted to engage in a business venture to house tourists who want to take a break from modern urban life and enjoy the feeling of returning to nature. For the founder, going back to nature meant minimizing his lifestyle and conserving nature along with catering to guests at BtN with organic home-based products. Thus, BtN has a vision to inspire other local enterprises to examine more sustainable and harmonious approaches by showing that they too can be successful. The business started in 2015 after 4 years of preparation. To reach BtN, visitors need to take a half-hour local bus ride along the bank of Phewa Lake from Hallan Chowk in Lakeside to Pame, which serves as the entry point of the World Peace Biodiversity Park (Fig. 5.2), and then must hike uphill for another half hour. The site was designed with the Satoyama concept in mind, with the rice field

Table 5.1 Basic information of the study area (prepared by authors)

Country	Nepal
Province	Gandaki
District	Kaski, Parbat, Syangja
Municipality	Pokhara Metropolitan City Ward no 23
Size of geographical area (hectare)	5776
Dominant ethnicity(ies), if appropriate	Gurung
Size of case study (hectare)	3.05
Dominant ethnicity in the project area	Gurung
Number of direct beneficiaries (people)	2966
Number of indirect beneficiaries (people)	27,482
Geographic coordinates (latitude, longitude)	28.234334 N, 83.787744 E

Fig. 5.2 The path to BtN and the entry point for the World Peace Biodiversity Park. (Photo Credit: Dambar Pun)

in the upper part of Phewa Lake with hills and the pristine forest landscapes (Fig. 5.4).

4 Community Initiatives for Sustainable Tourism

A 100-year-old sacred fig (*Ficus religiosa*, commonly called *peeple* in Nepali) and a 100-year-old cotton tree (*Bombax ceiba*, commonly called *simal* in Nepali) welcome guests in the front yard of the BtN along with a 50-year-old Hindu Shiva Temple. The homestays are located around the midsection of the Bhedikharka and Jhyapulokhe community forests (CFs) on the west side, and the World Peace Biodiversity Park (WPBP) is on the east (Pun et al. 2023). The Jauchare Dadhakarkha Bhirim CF, now known as WPBP, was originally established for the initiation of ecotourism with a focus on biodiversity conservation. The cessation of consumptive wood and fuelwood usage in WPBP serves to promote a sustainable tourism model between the Panchase Region (PR) and BtN. These activities are supported by the Community Forest Ecotourism Promotion Guideline of 2018 for promotion,

Fig. 5.4 View of Harpan River, the feeder of Phewa Lake, agricultural land, and Pokhara valley from BtN. (Photo Credit: Rebecca Gurung)

implementation, and monitoring of community-based ecotourism business (MoFE 2018). Support provided to ecotourism programmes/activities for private entities, communities, and cooperatives under the Forest Act of 2019 and Forest Regulation Act of 2021 further enhance benefit sharing between government and local community enterprises. From the northern side of the BtN, a spectacular range of Dhaulagiri, Machhapuchhre, Annapurna, and Manaslu Himalayas (Fig. 5.3) can be seen, with Phewa Lake and agricultural lands in the front view (Fig. 5.4).

There is a small river called the Kalchaura, which is named after the Himalayan whistling thrush, called *kalchuara* in Nepali. This river becomes another feeder and adds water to the Harpan River during monsoon season (June to September). These areas of BtN are to be included in the future expansion of the World Peace Biodiversity Park, Pokhara.

5 Nature-Positive Activities at BtN

There are a number of activities at Back to Nature (BtN) that contribute to achieving a nature-positive approach to ecotourism development. These include birdwatching, guided forest walks, forest restoration through nature education and interpretation, firefly conservation, agroforestry, watershed protection, and community forest management. The following activities reflect the focus of the BtN, which is to maintain nature conservation in all of its products and services. The information for this

Fig. 5.3 Earthen house built by BtN with snow-capped mountains in the background. (Photo Credit: Dambar Pun)

section was obtained from interviews with BtN owner and staff and focus group discussions on experiences and waste management at BtN.

5.1 Farm to Table Experience

Going back to nature is the *mantra* of BtN, thus the foods provided there are grown in the agricultural lands of the PR. As such, BtN grows its own vegetables and staple foods like rice on its own land by organic methods and buys other local products from the local communities. Between 2160 and 3600 kg of rice are produced on the 1.5 hectares of agricultural land as per the weather conditions of the planting year. Another 0.25 hectares are used for growing seasonal vegetables and spices like ginger, garlic, turmeric, etc., 0.25 hectares for livestock (two buffaloes, two cows, ten goats, 25–30 free range hens), and 0.51 hectares of grass and forest area are for the farm animals for manure and milk products) (BtN 2022). BtN owner and staff were interviewed about the agrobiodiversity, which includes "ten beehives and various fruit-bearing trees such as oranges, papaya, bananas, passion fruits, grapes, and various berry varieties, along with multiple useful grass fodders like *Thysanolaena maxima* (locally called *amrislo* in Nepali)" (refer to Fig. 5.5). A focus group discussion was conducted with communities in these regions, revealing that "BtN actively promotes the use of local rice varieties and seasonal vegetables. Consequently, the rice, identified as *Oryza sativa* (a variety called *jhinuwa* in Nepali), is both planted and harvested by local communities, and BtN utilizes the produce to serve guests."

5.2 Waste Management

BtN staff and owner were asked in semi-structured interviews to explain how degradable and non-degradable wastes are managed by the kitchen staff. Waste water from the kitchen is used in kitchen gardens, and biodegradable waste is converted into fertilizer used on vegetables. Similarly, manures from the farm are used in the agricultural lands for paddy farming. In addition, non-degradable wastes like plastics are sent to authorized collectors in the city. For other toilet wastes, septic tanks have been constructed. The wastes are recycled when necessary.

5.3 Types of Tourists and Length of Their Stay

BtN also runs an organic nature yoga class, bird watching, singing bowl therapy, jungle walks, and short hikes in and around the two CFs and WPBP aligned with BtN. International tourists usually stay for a week and may extend their stay up to a month, whereas domestic tourists spend 2 days on average. More tourists visited BtN annually after the pandemic for nature-based tourism (BtN 2022). The tourists can choose any accommodation type, that is, either tent houses (75-person capacity) or regular rooms (40 persons) in traditional houses constructed from local materials (BtN 2022). BtN believes in preserving vernacular houses and traditional crafts and has built earthen houses with local materials in line with greenery and

Fig. 5.5 Preparing a household broom from the *amrislo* fodder plant in PR. (Photo Credit: Rebecca Gurung)

environment-friendly waste management. It has abandoned pesticide use for animal and green manure practices on the farm and promotes a cognitive learning environment on nature, biodiversity, wildlife, culture, and tradition.

5.4 *Livelihood Diversification in Communities*

Direct interviews and focus group discussions with communities suggested there has been an increase in the number of restaurant and homestay services in the region after the launch of BtN and WPBP, with the involvement of two to three households in organic farming due to the expansion of the market from tourism. An eco-trail was constructed from Chapakot village to WPBP which connected both BtN and PPF. Trekking routes can provide prolonged tourist opportunities in these areas (Fig. 5.2). BtN is a new enterprise developed on the concept of nature-based tourism. Currently, nine people from different parts of Nepal (five are local) are employed by BtN and part-time employees are hired as required. A family looks after the farm and feed for the animals, three persons look after housekeeping, two work in the kitchen, and two serve as waiters. An incentive-based plantation programme with local people has been implemented by BtN in the WPBP and on adjoining abandoned agricultural land, which enhance biodiversity in these regions. Hence, livelihood activities in the communities that were previously subsistence farming associated with consumptive use of forest products like timber, fuelwood, NTFPs, soil, and stones, were transitioned into non-consumptive commercial businesses like restaurants, homestays, commercial organic farming, and employment opportunities as staff of BtN.

5.5 *Conservation Initiatives*

BtN has been an advocate for conserving cultural and traditional heritage and biodiversity. Hence it took special initiative to extend the World Peace Biodiversity Park to reach the Pumdi-Bhumdi passing land owned by BtN. BtN was one of the local actors who brought an end to the extension of a rural road to stop the destruction of biodiversity and avoid natural calamities like landslides and soil erosion, which are prime occurrences after road extension in rural areas. Hence, the BtN side of Chapakot of the PR has remained pristine and natural. Firefly tourism is promoted during the monsoon season as a new tourism product. BtN believes in preserving vernacular houses and traditional skills of crafts. Thus, its buildings were constructed in local traditional style using local materials (Fig. 5.6). BtN's earthen house made from local materials matches with the greenery around, providing a pollution-free pleasing visual. Use of pesticide is discouraged in BtN as well as surrounding communities. The conservation initiative has incorporated community-based conservation and ecotourism in the Annapurna Conservation Area (Gurung

2008), to enhance the governance structure with involvement of the provincial government, local government, BtN, and communities, presuming the sustainability of ecotourism and the conservation-based model within the PR (Pun et al. 2023). However, the traditional knowledge of local communities on herbs and medicine was discovered. Herbal tea and liquors, food, and nature tours have been provided to visitors by BtN. Fruit tree plantations and enhanced forest growth have created further opportunities for biodiversity in the WPBP region.

6 Discussion

This case study shows that BtN aims to establish itself as an ecotourism model, and that its successful management depends on ecosystem services provided by biodiversity (Pun et al. 2023). BtN is just an hour away from the major tourist city of Pokhara, where the primary purpose of tourists is to enjoy pristine nature and local culture. Especially in the picturesque PR where BtN is located, the conservation of biodiversity and culture has been the major interest of BtN and stakeholders of the CFs. To minimize the impacts of tourism and heavy dependency on forest resources, BtN provides foods from its own organic farm and agricultural lands, practises environment-friendly waste management, and economically supports more than nine families with the various conservation initiatives it has undertaken since 2011. BtN, which is a prime beneficiary of the WPBP established in this area along with a newly formed restaurant and homestays, needs an effective means to establish a benefit-sharing mechanism focusing on conservation incentives. Since 2020 with the initiation of WPBP management plan activities, BtN has financially as well as

Fig. 5.6 Traditional house at BtN. (Photo Credit: Dambar Pun)

socially collaborated with communities and government organizations for implementation and monitoring of interventions (Pun et al. 2023). For the success of community-based conservation initiatives, the benefits of conservation must address the issues faced by communities through their participation in decision making and/or incentive-based measures whereby local communities can receive some benefits (Bajracharya et al. 2005; Snyman 2013).

The WPBP management plan has an annual budget set aside for ecotourism and biodiversity initiatives up to 2024 (Pun et al. 2023). After the first year of implementation in 2019, COVID-19 triggered political instability. The lack of participation due to financial constraints imposed by the government (Panchase Protected Forest Section) slowed the progress of the management plan in subsequent years. The full implementation of the WPBP management plan with communities' self-motivated interest in ecotourism initiation in these areas can be replicated from the Annapurna Conservation Area (e.g. Bajracharya et al. 2005) for the success of ecotourism and biodiversity conservation initiatives. External actors or donor agencies like non-governmental organizations and the private sector can assist in closing financial gaps as well as providing technical support in community-based resource management or ecotourism projects.

Further, the vision of BtN concurs with the Satoyama concept. BtN cannot survive without being in harmony in nature, which is the vision of the Satoyama concept (IPSI 2023). BtN is an advocate of the preservation of traditional knowledge and skills, farm-to-table experiences, and conservation of nature. It has taken the initiative to develop the WPBP and is setting an example by saving natural forests from potential disasters caused by unplanned rural road construction. Therefore, its practices align with the threefold approach of the Satoyama Initiative (IPSI 2023).

7 Conclusions

The case study shows that a nature-positive approach to promoting a sustainable ecotourism business model can achieve biodiversity conservation based on ecosystem restoration and a participatory approach involving communities and private organizations like BtN. Biodiversity conservation and livelihood enhancement can be achieved by an enhanced governance structure based on an effective benefit-sharing mechanism between communities and the private sector, as BtN based its ecotourism practice on the Community Forest Ecotourism Promotion Guideline with traditional knowledge used for incentive-based conservation (MoFE 2018). Nature-positive approaches do not need to be complex or technical. Simply implementing traditional knowledge to secure food for the future, using terraced farming in hilly areas, conserving skills related to water sources, and the selling of non-timber products from community/personal forests can be nature-positive activities. A nature-positive approach, as well as ecotourism, gives emphasis to ecosystem restoration with integrated soil, water, forest and other natural resources and places importance on socio-economic enhancement with livelihood diversification from employment and agricultural opportunities, enterprise development, and tourism. Policy interventions and activities require synergies. Support from external actors, either financially, technically, or cognitively, can produce sustainability for the WPBP with community-based conservation interventions. These good practices can be replicated in other private tourism sectors as a socially and ecologically balanced sustainable tourism business model for ecosystem restoration integrating "Nature Positive, Ecotourism and SEPLs" approaches. Integration of these approaches can create a positive environment for sustainable tourism and a model for tourism entrepreneurs leading to the utmost preservation and protection of critical habitats in the Panchase Region as well as in Nepal.

References

Adhikari JN, Bhattarai BP, Thapa TB (2018a) Human-wild mammal conflict in a human dominated midhill landscape: a case study from Panchase Area in Chitwan Annapurna Landscape, Nepal. Journal of Institute of Science and Technology 23:30–38, viewed 1 February 2023. Retrieved from https://www.nepjol.info/index.php/JIST/article/view/22158

Adhikari S, Baral H, Nitschke C (2018b) Adaptation to climate change in Panchase Mountain Ecological Regions of Nepal. Environments 5(3):42, viewed 11 February 2023. Retrieved from https://www.mdpi.com/2076-3298/5/3/42

Anup KC (2017) Ecotourism in Nepal. The Gaze Journal of Tourism and Hospitality 8(1):1–19, viewed 9 February 2023. Retrieved from https://www.nepjol.info/index.php/GAZE/article/view/17827

Back to Nature (BtN) (2022) Annual report and registration, Pokhara, Nepal

Bajracharya SB, Furley PA, Newton AC (2005) Effectiveness of community involvement in delivering conservation benefits to the Annapurna Conservation Area, Nepal. Environmental Conservation 32(3):239–247

Bhandari AR (2020) Assessment of ecosystem services on Panchase Protected Forest, Nepal. Ph.D. thesis, Central Department of Environment Science, Tribhuvan University, viewed 17 February 2023. Retrieved from https://elibrary.tucl.edu.np/handle/123456789/9553

Bhandari P, Budhamagar S, Shrestha KK (2018) A checklist of flowering plants of Panchase Protected Forest, Kaski district, central Nepal. Journal of Natural History Museum 30:55–84. Retrieved from https://www.nepjol.info/index.php/JNHM/article/view/27538

Buckley R (1994) A framework for ecotourism. Annals of Tourism Research 21(3):661–669

Buckley R (2011) Tourism and environment. Annual Review of environment and Resources 36:397–416, viewed 27 February 2023. Retrieved from https://www.annualreviews.org/doi/abs/10.1146/annurev-environ-041210-132637

Bull JW, Milner-Gulland EJ, Addison PF, Arlidge WN, Baker J, Brooks TM, Burgass MJ, Hinsley A, Maron M, Robinson JG (2020) Net positive outcomes for nature. Nature Ecology & Evolution 4(1):4–7

Davlatbek X, Doniyorbek X (2023) Further improvement of tourism in our country through development of ecotourism in Uzbekistan. British View 8(1):3–6, viewed 29 January 2023. Retrieved from https://www.britishview.co.uk/index.php/bv/article/view/167

Forest Act (2019) Ministry of Forest and Environment, Nepal

Forest Regulation Act (2021) Ministry of Forest and Environment, Nepal

Government of Nepal, Ministry of Culture, Tourism & Civil Aviation (MoCTCA) (2020) Nepal tourism statistics 2020, Singhadurbar, Kathmandu, Nepal, viewed 30 January 2023. Retrieved from https://www.tourism.gov.np/files/NOTICE%20MANAGER_FILES/NTB_Statistics_2020%20_final.pdf

Government of Nepal, Ministry of Culture, Tourism & Civil Aviation (MoCTCA) (2022) Nepal tourism statistics 2022, Singhadurbar, Kathmandu, Nepal, viewed 30 January 2023. Retrieved from https://www.tourism.gov.np/files/NOTICE%20MANAGER_FILES/Setting_Nepal%20 Tourism%20Statistic_2022.pdf

Gurung HB (2008) Innovation in Ecotourism: Its Contribution to Community-based Conservation. In: *CAUTHE 2008: Tourism and Hospitality Research, Training and Practice; "Where the 'Bloody Hell' Are We?"*, proceedings of the 18th Annual Council for Australian University Tourism and Hospitality Education (CAUTHE) conference, 11–14 February 2008, Gold Coast, Qld, Griffith University, pp. 197–202

International Partnership for the Satoyama Initiative (IPSI) (2023) The Satoyama Initiative, viewed 19 February 2023. Retrieved from https://satoyama-initiative.org/concept/satoyama-initiative/

Locke H, Rockström J, Bakker P, Bapna M, Gough M, Lambertini M, Morris J, Polman P, Rodriguez CM, Samper C, Sanjayan M, Zabey E, Zurita P (2021) A Nature-Positive World: The Global Goal for Nature. https://f.hubspotusercontent20.net/hubfs/4783129/Nature%20 Positive%20The%20Global%20Goal%20for%20Nature%20paper.pdf

Ministry of Forests and Environment (MoFE) (2018) Community Forest Ecotourism Promotion Guideline. Ministry of Forest and Environment, Government of Nepal. https://dforautahat.gov.np/wp-content/uploads/2022/07/Community-forest-Karyabidhi_tourism_2075.pdf

Nyaupane GP, Thapa B (2004) Evaluation of ecotourism: a comparative assessment in the Annapurna Conservation Area Project, Nepal. Journal of Ecotourism 3(1):20–45, viewed 14 February 2023. Retrieved from https://www.tandfonline.com/doi/abs/10.1080/14724040408668148

Ojha N (2020) Potential of ecotourism in Nepal. Interdisciplinary Journal of Management and Social Sciences 1(1):104–112, viewed 10 February 2023. Retrieved from https://www.academia.edu/67401834/Potential_for_Ecotourism_in_Nepal

Paudel DP (2017) Forest resource management and livelihood of local people in Panchase Area of Western hilly region, Nepal. The Third Pole: Journal of Geography 17:37–50, viewed 15 February 2023. Retrieved from https://www.nepjol.info/index.php/TTP/article/view/19981

Pörtner H, Scholes R, Agard J, Archer E, Arneth A, Bai X, Barnes D, Burrows M, Chan L, Cheung W (2021) IPBES-IPCC co-sponsored workshop report on biodiversity and climate change. IPBES and IPCC, 10.

Pun D, Tiwari A, Shrestha P, Dhungana K, Gahatraj G, Pun DBP (2023) Initiation of SEPLS Approach from World Peace Biodiversity Park (WPBP), Pokhara in Panchase Region of Nepal. In: Nishi M, Subramanian SM (eds) Ecosystem Restoration through Managing Socio-Ecological Production Landscapes and Seascapes (SEPLS). Springer Nature, Singapore, pp. 61–75

Regmi KD, Walter P (2017) Modernisation theory, ecotourism policy, and sustainable development for poor countries of the global South: Perspectives from Nepal. International Journal of Sustainable Development & World Ecology 24(1):1–14

Samal R, Dash M (2023) Ecotourism, biodiversity conservation, and local livelihoods: Understanding the convergence and divergence. International Journal of Geoheritage and Parks 11(1):1–20. https://doi.org/10.1016/j.ijgeop.2022.11.001

Sen A, Dabi N (2021) Tightening the Net: Net zero climate targets – implications for land and food equity. Oxfam. https://doi.org/10.21201/2021.7796

Simmons DG (2013) Tourism and Ecosystem Services in New Zealand. In: Dymond JR (ed) Ecosystem services in New Zealand – conditions and trends. Manaaki Whenua Press, Lincoln, New Zealand, pp. 343–348

Snyman S (2013) Household spending patterns and flow of ecotourism income into communities around Liwonde National Park, Malawi. Development Southern Africa 30(4–5):640–658

Sonter LJ, Dade MC, Watson JE, Valenta RK (2020) Renewable energy production will exacerbate mining threats to biodiversity. Nature Communications 11(1):1–6

United Nations World Tourism Organisation (UNWTO) (2023) Ecotourism and protected areas, viewed 10 February 2023. Retrieved from https://www.unwto.org/sustainable-development/ecotourism-and-protected-areas

WBCSD (2021a) Advancing business understanding of "nature positive." World Business Council for Sustainable Development. https://www.wbcsd.org/a9fvw

WBCSD (2021b) Practitioner guide - What does nature-positive mean for business? Geneva, Beijing, Delhi, London, New York, Singapore, World Business Council for Sustainable Development, viewed 31 January 2023. Retrieved from https://www.wbcsd.org/contentwbc/download/13439/196253/1

Yogi HN (2010) Eco-tourism and sustainability – opportunities and challenges in the case of Nepal. Master's thesis, Department of Sustainable Development, University of Uppsala, viewed 8 February 2023. Retrieved from https://www.diva-portal.org/smash/get/diva2:408751/FULLTEXT01.pdf

Dambar Pun CEO of Back to Nature, Nepal. Has worked on conservation and ecotourism-based business since 2003. Has conducted philanthropic work around the Panchase Protected Forest for community awareness and skill development on biodiversity conservation and ecotourism since 2011.

Aashish Tiwari Works at the Conservation Development Foundation, Nepal. Enthusiastic on conservation-based natural resource management. Has a Master's degree in forest management and biodiversity conservation from the Institute of Forestry, Pokhara, Nepal.

Rebecca Gurung Sustainability Officer involved with the Himalayan Sustainable Future Foundation. Has a Master's in environment management from Lincoln University, New Zealand. She has been working in the field of biodiversity conservation and climate change in Western and Mid-Western Nepal.

Hum Bahadur Gurung Asia Partnership Manager at BirdLife International Asia. He has a PhD from Griffith University, Australia. His research interests include participatory action and community-based approaches for environmental education, sustainable tourism management, and ecosystem services.

Samuel Pun Serves as Managing Director at Backyard by Back to Nature, a restaurant specializing in promoting eco-friendly and sustainable living practices, while studying at Westcliff University. Dedicated to sustainability, he contributes to preserving the environment and promoting a greener lifestyle.

BY NC SA

Chapter 6
Business-Culture-Biodiversity Nexus: The Foundation of Socio-Ecological Integrity in Traditional Cultural Landscapes in Indian Himalaya

K. G. Saxena, S. Sreekesh, K. S. Rao, R. K. Maikhuri, and S. Nautiyal

Abstract The multi-functionality of landscapes sculpted by Hindu, Buddhist, and Animist communities in tropical, temperate, and alpine biomes in Indian Himalaya was analyzed using remote sensing, ground sampling, and participatory/rural appraisal tools. All across, traditional people articulated business as common and confined interlinked "professions." Cultural values favoured equity, environmental sustainability, and social integration. Humanitarian, philanthropic, and altruistic motives prevailed over profits in exchanges. Profit-centric business appeared in the late nineteenth century with the colonial forest policy of revenue earning from timber previously utilized by people only for subsistence due to cultural/knowledge barriers. Agricultural business enterprises emerged after 1970 with the launch of national policies subsidizing inputs and securing staple food and the fast growing market of organic food, herbal medicine, and other non-timber traditional products. Amidst widespread negative impacts, there are a few landscapes testifying to the positive impacts of the market on biodiversity. Scientific improvement in traditional manure production and agroforestry, adoption of vermitechnology, rehabilitation of degraded lands, people-led ecotourism, and cooperative marketing/value addition can promote business coupled with biodiversity conservation. It is concluded that (i)

K. G. Saxena (✉)
School of Environmental Sciences, Jawaharlal Nehru University, New Delhi, India

S. Sreekesh
Centre for the Study of Regional Development, School of Social Sciences, Jawaharlal Nehru University, New Delhi, India

K. S. Rao
Department of Botany, University of Delhi, Delhi, India

R. K. Maikhuri
Department of Environmental Sciences, HNB Garhwal University, Srinagar, Uttarakhand, India

S. Nautiyal
Centre for Ecological Economics and Natural Resources, Institute for Social and Economic Change, Bengaluru, Karnataka, India

M. Nishi et al. (eds.), *Business and Biodiversity*, Satoyama Initiative Thematic Review, https://doi.org/10.1007/978-981-97-7574-3_6

enhancing local income from value addition/marketing is now crucial for conserving forest biodiversity, (ii) people need to be clearly informed of business/market risks and uncertainties, (iii) value addition and marketing should be included in community forestry currently confined to restoration/protection, and (iv) there is a need of participatory research to uncover the science underlying traditional biodiversity/land uses and prospective market products, and to incorporate this information in People's Biodiversity Registers.

Keywords Invasive species · Medicinal herbs · Restoration · Traditional knowledge · UN-REDD · Value addition

1 Introduction

Despite overwhelming positive trends in supportive policies and actions, the world failed to meet the Aichi Targets aimed at preventing and reversing the loss of biodiversity and ecosystem functions (Secretariat of Convention on Biological Diversity 2020). This failure has brought out a need to bolster the participation of business enterprises that goes beyond existing corporate social responsibility mechanisms and voluntary adoption of the Principles for Responsible Investment to promote opportunities for the private sector in biodiversity actions (Lambooy and Levashova 2011). Virtually, all industrial and business enterprises cause some loss of biodiversity, while the ones utilizing biomass/organisms risk collapse due to ecological degradation. Thus, environmental conservation, restoration, compensation, and reparation deserve as much importance in business development plans as physical and financial growth (Panwar et al. 2022). Business, in common parlance, means professional activities and enterprises or institutions/organizations investing in the production of knowledge, goods, and services for profit while consciously taking some risks. Hence, business is intrinsic in the functioning of not only private industrial/business enterprises but also of governmental/non-governmental organizations, communities, households, and even individuals.

The nature and determinants of biodiversity and ecosystem functions are such that their conservation and restoration demand synergy among different stakeholders. While consolidating conservation efforts is a global imperative, biodiversity hotspots deserve priority for hosting more than 50% of world's endemic plant species and 43% of mammal, bird, amphibian, and reptile species, and delivering 35% of ecosystem services from just 2.5% of the Earth's surface they cover (Mittermeier et al. 2004). The Himalayan hotspot is immensely rich in both wild and domesticated biodiversity, and the ecosystem services it provides support the well-being of 50 million people living within this mountain system, as well as 2.5 billion outside of it in two megadiverse countries (China and India) and six other countries (Afghanistan, Bangladesh, Bhutan, Myanmar, Nepal, and Pakistan) of Asia.

Efforts have been made to understand the spatio-temporal dynamics of biodiversity in large management units (Wester et al. 2019). However, there is a dearth of comprehensive efforts on understanding these dynamics and their linkages with business in traditional socio-ecological production landscapes or village landscapes—the basic units of sustainable development planning and programme implementation. These ancient landscapes, isolated by terrain, linguistic, and cultural barriers and shaped by a subsistence economy, began to be exposed to the cash market/industrial economy from the nineteenth century alongside modern infrastructure development driven by national governments. Thus, comparisons of landscapes differing in accessibility but inhabited by the same ethnic/cultural group unfold the socio-ecological impacts of the cash market economy and conventional businesses. Likewise, the convergence/divergence between different ethnic/cultural groups are uncovered by comparing landscapes with similar accessibility and biophysical conditions managed by them. This article is an attempt to (i) analyze biodiversity-culture-business interlinkages in village landscapes differing in accessibility, culture/ethnicity, and biophysical conditions, and (ii) explore business options for conservation and sustainable utilization of biodiversity in the Indian Himalayan Region.

2 Materials and Methods

Traditional villages in Indian Himalaya are ancient socio-ecological production landscape mosaics of farms and forests scattered around vast government-owned forest/alpine areas free from agriculture that were carved out in the late nineteenth century. Village communities are fully empowered to manage their respective territories, excepting non-forest uses of forest land and commercial utilization of forest resources. Further, the people have been granted free customary subsistence resource utilization in government forests/alpine areas, but access is regulated by the Forest Department at the respective provincial levels.

Shifting agriculture is the predominant food production system of the nature-worshipping ethnic minorities in the subtropical humid climate of eastern Himalaya, while settled sole-crop/tree-crop mixed agriculture is practised by the Hindu/Bhotiya[1] of Aryan origin in the temperate/subtropical moist climate of central Himalaya, and Buddhists with Tibetan affinity in cold arid northwestern Himalaya engage in an agro-pastoral system. Two village landscapes, one typical of the most accessible and the other the least accessible, were selected in each region. Farmland holdings were ascertained from village records. The forest and pasture areas available to an average household were obtained by dividing village forests and pasture areas ascertained by superimposing the segments delimited by natural boundaries shown by people onto digitally classified Landsat/Indian Remote Sensing Satellite data by number of households. Google Maps served as ancillary data for checking

[1] Bhotiya is an exonym for people living around alpine regions and securing livelihoods from pastoralism and trade with Chinese and Tibetans through mountain passes.

land cover. Flowering plant diversity and productivity (harvest method) were assessed in 40 × 20 m plots (n = 5 in each land cover/use type in each village). Social stratification by profession/business, monetary values of produce, supply chain, price spread, and farmers' perceptions about biodiversity, industry/business, and culture were discerned from rapid rural appraisals, followed by household surveys and participatory discussions with farmers, traders, officials of relevant government departments, cooperatives, and companies. Efforts were made to get an account of both the existing and past states of business and biodiversity as recalled by the people.

The study covered social-economic-ecological attributes of biodiversity viz. land use/cover, ecosystem types, number of crop species/cultivars, livestock species, wild vascular plant species, invasive species, and animals attracting tourists. The basal area of trees was used as a surrogate for carbon mitigation potential. Information on policies since the colonial regime was compiled, analyzed, and compared with the ground scenarios over the 1990–2015 period (see Saxena et al. 2011; Shimrah et al. 2015; Rao et al. 2022a, b for details). This intensive study of six villages was complemented by information gathered from rapid appraisals and participatory discussions in nine other villages (Fig. 6.1, Table 6.1).

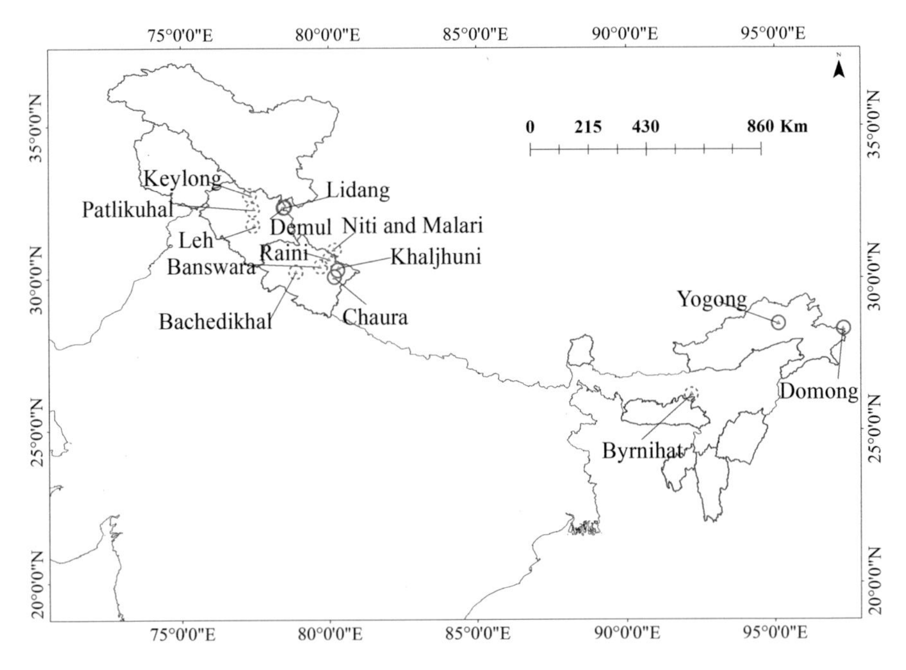

Fig. 6.1 Location of six landscapes selected for the intensive long-term study (viz. Domong, Yogong, Chaura, Khaljhuni, Lidang, and Demul) and others for rapid short-term studies. (Source: Google Earth Image/information extracted from Google Earth Pro software)

Table 6.1 Basic information of the study area

Study villages	Lidang	Demul	Chaura	Khaljhuni	Domong	Yogong
District	Lahaul and Spiti		Bageshwar		West Siang	
Province	Himachal Pradesh		Uttarakhand		Arunachal Pradesh	
Size of geographical area	Indian Himalayan Region: 54 million ha					
Dominant ethnicity	Spitians with faith in Buddhism and affinity with Tibetans		Hindus and Bhotiyas		Adis: Animists (nature worshippers)	
Elevation (m.a.s.l.)	3300–3500	4000–4500	1200–1300	2200–2400	1600-1800	1600–1800
Climate	Cold arid		Moist temperate		Humid subtropical	
Agricultural system	Agro-pastoral		Settled agroforestry		Shifting agriculture on slopes and wet paddy in flat valleys	
Accessibility (distance to motorable road in km)	Roadside	14	4	14	Roadside	14
Size of case study/project area (ha)	58	613	1759	216	4900	7622
Number of direct beneficiaries (people around intensively studied villages)	55	300	510	240	1759	216
Number of indirect beneficiaries (people of similar villages)	5000	10,000	20,000	5000	10,000	2000
Geographic coordinates	32°14'N, 78°14'E	32°17'N, 78°18'E	30°05'N, 79°94'E	30°11'N, 79°98'E	28°17'N, 97°04'E	28°33'N, 94°85'E

3 Results

3.1 *Traditional Perspective of Biodiversity*

Across the ethnic groups and bioclimatic regions covered in the study, farmers understood landscapes as mosaics of interacting land use/cover types differing in vegetation structure, species, cultivars, products, intangible functions, land/labour productivity, and resistance/resilience to climatic variability/uncertainty. Biophysical conditions were such that only one crop could be harvested each year in the cold desert and two in the moist temperate and humid subtropical climates. Cropping patterns in settled agricultural landscapes were built around the two seasons (rainy/summer and winter) unlike shifting agricultural landscapes where climate was quite even, or cold desert where winters were too cold to grow two crops a year. Homegardens, common to all landscapes, were the most productive (biologically but not necessarily monetarily), species-rich, resilient, and structurally complex agroecosystem that accomplished the dual functions of commodity production and waste recycling (Fig. 6.2a, b).

Resilience in the face of biophysical uncertainties was attempted by maintaining high cultivar/genetic diversity in irrigated fields distinguished by paddy monocropping in subtropical humid areas, wheat and barley in cold desert, and all three in the moist temperate climate, and by rotation/intermixing of numerous crops in other agro-ecosystems (Table 6.2).

Farmers classified crops as (a) highly remunerative and highly stress sensitive: maize, soybean, paddy, wheat, lentil, potato, buckwheat, amaranths, green pea, and vegetables; (b) less remunerative but highly stress tolerant: finger millet, foxtail millet, barnyard millet, and barley; and (c) highly remunerative and highly stress

Fig. 6.2 (**a**) Homegarden (rear) and crop fields in subtropical humid Himalaya. (Photo by authors) (**b**) Homegardens (small tree clad patches) in a cold desert village, Himachal Pradesh, India. (Photo by authors)

Fig. 6.2 (continued)

tolerant: sesame, cowpea, black pea (*Pisumarvense*), horse gram, and pigeon pea. Maximization of the area of *Panicum miliaceum* (considered as low-quality "rice") that matures in just 3 months was a widespread adjustment in unfavourable climate years in subtropical/temperate moist zones. For farmers, crop biodiversity was a resource to optimize returns from land and labour, and forest biodiversity a means to secure a wide range of free goods and services crucial for survival in isolated settlements around a dissected fragile terrain (Rao et al. 2022b).

Only in settled agricultural landscapes, farmers recalled famine resulting from drought in 1966–1967. They had to consume wild *Pyrus pashia* and *Aesculus indica* seeds as a staple food and use oil extracted from the seeds of *Princepia utilis* for healthcare. Likewise, they used the seeds of *Prunus americana* and *Litseolea consimilis* for cooking, as well as collected 15–20 wild species as food supplements and exchanged handicrafts for food grains with nomadic pastoralists. Religious/cultural beliefs restrained the killing of wildlife for food/income. Farmers understood that the maintenance of hydrological balance, and shielding of settlements from wind and protection from wildlife were the regulating ecosystem services of the forest (Rao et al. 2022a, b).

Table 6.2 Distinguishing features of ecosystem types in traditional socio-ecological production landscapes

Type of agriculture/ecosystem	Biodiversity[a]	Other important attributes
Homegardens	High crop species, functional, and structural diversity, but low genetic diversity	Nutritional security and potential of some cash income round the year, productive sink of household wastes, low impact of climatic variability and uncertainty, high land and labour productivity[b]
Settled rainfed agroforestry/shifting agriculture	High crop species, functional, and genetic diversity, but medium structural diversity	Food diversity and security, supplementation of forest products from agroforestry in settled and fallow vegetation in shifting agriculture, low impact of high rainfall regimes, low land and high labour productivity[b]
Irrigated sole-crop system	Low crop species, functional, and structural diversity, but high genetic diversity	Staple food security, low impact of low rainfall regime, scope of growing both local food and cash crops, high land productivity and low labour productivity[b]
Forests/meadows	Highest levels of structural, species, genetic, and functional diversity	Availability of free agricultural inputs and a range of other products required for subsistence, maintenance of hydrological balance, cultural services for supplementary income

[a]Crop diversity: number of crop species; Crop genetic diversity: number of cultivars; Functional diversity: representation of cereals, millets, beans, oil seeds, and green vegetables in crops grown and of shade tolerant/intolerant, evergreen/deciduous, nitrogen-fixing/non-fixing, shallow/deep-rooted species, fire sensitive/resistant species in forests; High, medium, and low structural diversity: three, two, and one vertically stratified layers of vegetation
[b]Land productivity: economic yield (kg)/ha/year; Labour productivity: economic yield (kg)/person/day of labour input

3.2 *Traditional Perspective of Business and Trade*

Deductions from participatory discussions supported by transactions observed in the field revealed four levels of business/trade operations: (i) within a village community; (ii) between village communities, nomadic pastoralists (shuttling between their summer settlement in alpine meadows and winter settlement in foothills), and Bhotiya people (shuttling between their main dwelling in cool temperate and satellite dwelling in warm temperate region); (iii) between tourists and village communities; and (iv) cross-border trade between Indian Bhotiyas and high-altitude Tibetans and Chinese nationals.

Across all types, ploughing, herding, milling, leatherworking, metal/woodworking, therapy, priesthood, and trading outside native villages were specialized professional activities of families with small landholdings. Wool-making, spinning, dyeing, weaving, crafting, food processing, portering, guiding, and farm operations other than ploughing were common to all families. Men performed more labour-intensive and risky tasks (e.g. ploughing, nomadic pastoralism, cutting/processing/transporting bamboo/timber, collecting lichens from tree canopies, portering, and

Table 6.3 Traditional beliefs, norms, and practices favouring environmental sustainability, equity, and social integration

Beliefs, norms, and practices	Favoured attribute(s)
Religious beliefs: (i) cutting green trees and killing wildlife for income and abandonment of settled farming invite divine curses on the entire community, (ii) pruning/lopping lead shoots of fodder/fuelwood trees invite divine curses on the concerned individuals and their families, and (iii) religious rituals can check the losses/enable fast recovery in the face of extreme climate events	Environmental sustainability
Utilization of forest products only by residents in groups constituted in open village assemblies as a mechanism of facing the risks from wildlife, checking poaching, and preventing overexploitation	Environmental sustainability
Rotational leaf litter removal, removal of other non-timber products, and grazing by livestock in transit from/to alpine meadows is practised in all forests except sacred forests. Rotation period and utilization intensity is decided based on the current ecological state	Environmental sustainability
All households have equal rights in natural forests/meadows but only small landholders can sell handicrafts	Equity
Only local healers can store rare medicinal herbs	Environmental sustainability
Labour from outside the village can be hired only after exhausting all opportunities within and after seeking consent of the whole community	Equity and environmental sustainability
Only smallholders can earn income from ploughing, carpentry, smithery, leatherworks, tailoring, milling, and herding	Equity
Nomadic pastoralists (the landless minorities) coming from foothills are allowed free grazing in return for their livestock shielding local livestock from wildlife and suppressing dominance of uneconomic species in distant areas not grazed by local livestock. These nomads exchange commodities not available locally (fabric, jiggery, and other manufactured items) for local products (food grains) demanded by Tibetans	Social integration and environmental sustainability
Division of geographical area among local priests with mutual consent for conducting religious tourism/pilgrimages to shrines headed by priests from far-off places[a] or among village communities for treks to summits/glaciers; viewing pedestrian tourists/pilgrims as embodiments of Gods/Goddesses.	Social integration and equity

[a]Only in cold desert and moist temperate areas

trading outside the village), while women engaged in less labour-intensive and lower-risk ones (e.g. sowing, weeding, grass-cutting, fuelwood/fodder collection, and farmyard manure production). Social sanction of income from non-timber forest products (NTFPs) and technical tasks like carpentry, herding, and leatherworks going only to weaker families favoured equity (Table 6.3).

Traditions and beliefs favoured environmental sustainability and social integration. Participatory discussions revealed that business was founded on trust, respect, philanthropy, and altruism. Thus, people did not question nomadic pastoralists on the natural mortality/depredation of livestock under their custody during summer in alpine areas. The rich tended to pay higher than market prices for products/services sought from the poor. Profits were immaterial in the exchange of seeds or knowledge. People viewed providing basic facilities to pedestrian pilgrims/trekkers as their moral duty and not just a means of income generation. The traditional kingdom continues as the owner of shrines/sacred natural sites and people's institutions as the authority to spend donations for societal works.

People acknowledged enrichment of knowledge as a non-material benefit from pilgrims, tourists, and trekkers. Bhotiyas and nomadic pastoralist natives of alpine and foothill areas, respectively, were considered the most proficient traders. Bhotiyas exchanged food grains for salt, wool, medicinal/aromatic herbs, and gold with Chinese people and nomadic pastoralists, but they exchanged jaggery, fabric, and other consumer goods for food grains and potato only with locals. The people viewed nomadic pastoralism as a cost-effective way of checking weeds and crop/livestock depredation by wildlife. Parallels of credit, interest, tax, insurance, and marketing cooperatives were altogether lacking in the traditional socio-economic systems.

3.3 *Policy Interventions Influencing Biodiversity and Business*

The social mechanisms favouring biodiversity conservation, equity, and social integrity started declining after 1870s when the government took over all uncultivated lands and leased land to tea companies. In the early twentieth century, the government pronounced a policy of revenue generation from resources that the local people utilized only for subsistence, according to religious/cultural restraints (timber), or did not utilize at all for lack of knowledge (resin/charcoal). Government agencies fetched a huge income from logging industries, but neglected the needed investments in restoration and conservation. This business tailored for industrial development resulted in scarcity of the non-timber products crucial for local livelihoods. This in turn led to mass movements that drove policy changes: formulation of a new category of forests (community forests carved out from government forests) in the 1920s; prohibition of logging, hunting, and non-forest uses of forest land; creation of protected areas; and explicit recognition of the social and ecological functions of forests as more important than industrial and economic functions in the 1980s and 1990s. These radical policy changes coincided with the economic liberalization set out in 1992. Policy interventions over the 1990–2010 period addressed modalities of people's participation in forest management, financial support for organic farming and cultivation of rare/threatened medicinal herb species, rewards for maintaining vast and healthy forest cover (Green Bonuses: funds distributed among states by the national government in proportion to their contribution to national forest cover and ecosystem services), and compensatory afforestation in lieu of deforestation for essential services (roads, electricity, water supply, modern healthcare, education, and communication). The post-2010 period saw additional policy interventions targeting participation of business enterprises in environmental management, protection of people from exploitation in the market and natural calamities, production of forest commodities on private farms, and increasing investments (Green Bonds: funds raised by the national government from fixed-income investors to fund ecological restoration activities).

3.4 Biodiversity and Business in Village-Level Socio-Ecological Production Landscapes

Landscapes differed more in the land uses constituting them, species composition, and material inputs/outputs than in species richness (Table 6.4A, B). In all three regions, less-accessible landscapes had larger landholdings, and homegardens were the most species-rich but smallest land use type, while crop husbandry was economically a more efficient system than livestock husbandry and collection of non-timber products from forests. In shifting agricultural landscapes, wet paddy cultivation had the lowest species richness but highest productivity, while Toko palm plantations were the most extensive, but banana plantations the most remunerative land use (Table 6.4C). In settled agricultural landscapes, paddy/amaranth was the most productive and finger millet the most extensive crop in the rainy season. Further, naked barley was the most extensive and remunerative crop in Khaljhuni, unlike Chaura where barley was the most remunerative but wheat the most extensive crop. People earned income from tourism only in the cold desert region and from non-timber products from natural forests only in subtropical/temperate regions.

Time series data for all villages could not be created, but changes over the 1990–2015 period could be deduced from observations over frequent field visits, rapid appraisals, and participatory discussions with the communities. Deforestation or forest degradation did not occur in any landscape. Three dramatic changes in existing plant biodiversity were noted: (i) fruit tree planting in fallows in more accessible shifting agricultural landscapes, (ii) establishment of some exotic/invasive species in more accessible shifting agricultural landscapes, and (iii) cultivation of sweat pea in both cold desert agro-pastoral landscapes. The prominent changes in species distribution and composition included: voluntary planting of around five economic forest tree species in the Domong shifting agricultural landscape; planting of six economic herbs as part of a participatory restoration project in the Khaljhuni settled agricultural landscape; and substantial replacement of staple with cash crops in all settlements. Tourism business was confined to cold desert landscapes because of their uniqueness, and emerged after they became accessible by road in 2003. The highest permanent settlements in the world, fair chances of sighting blue sheep, wolves, and snow leopards, traditional methods of checking human-wildlife conflicts, and Buddhist shrines were the main attractions for tourists. Abandonment of water mills soon after electrification was a case of termination of a traditional business over the past few decades. People in settled agricultural landscapes used to earn from treks to glaciers/summits and local markets around

Table 6.4 Social-economic-ecological attributes of village landscapes selected for intensive studies in Indian Himalaya

	Agro-pastoral landscapes in cold desert		Settled agroforestry landscapes in moist temperate climate		Shifting agricultural landscapes in humid subtropical climate	
	Lidang	Demul	Chaura	Khaljhuni	Domong	Yogong
(A) **Land use/cover (ha/household)**						
Settled rainfed cropland	0.55	0.35	1.21	2.34	–	–
Shifting agriculture-cropped land	–	–	–	–	0.6	0.2
Shifting agriculture-fallows	–	–	–	–	4.78	3
Plantations	–	–	–	–	1.65	–
Homegardens	<0.01	<0.01	<0.01	<0.01	0.01	0.01
Wet paddy in valleys	–	–	–	–	0.15	0.02
Pasture-hay area	0.68	3.45	–	–	–	–
Pasture-for grazing by livestock of the village	1.59	1	–	–	–	–
Pasture-for grazing livestock of a village cluster	–	6.3	–	–	–	–
Forests degraded (canopy cover <30%)	–	–	3.2	6.3	–	–
Forests dense (canopy cover >70%	–	–	4.3	7.8	9.54	208
Total land (ha/household)	2.82	11.1	8.71	16.35	14.32	211
(B) **Agro-biodiversity**						
Livestock (number of animals/household)	2.6	10.6	35.5	29.1	4	4
Crop species richness (number of crops grown/household)	7	7	14	19	20	20
Crop cultivar richness (number of cultivars grown/household)	–	–	10	12	18	24
Livestock diversity (number of livestock species/household)	7	7	6	6	2	2
Wild plant species richness	80	84	70	70	92	96
Average basal area (m^2/ha)	–	–	64	81	74	84

(continued)

Table 6.4 (continued)

	Agro-pastoral landscapes in cold desert		Settled agroforestry landscapes in moist temperate climate		Shifting agricultural landscapes in humid subtropical climate	
	Lidang	Demul	Chaura	Khaljhuni	Domong	Yogong
Invasive species	None	None	None	None	*Eupatorium* sps. *Ageratrum* sps. *Mikania* sps.	None
Wildlife/livestock of tourist attraction	Yak, Dzo/ Dzomo	Blue sheep, Yak, Dzo/ Dzomo	None	None	Mithun	Mithun
(C) **Household income (INR/household)**						
Total household income (INR)	69,250	43,250	3380	4450	25,890	10,155
Monetary value of all marketable crops produces (% sold to market)	78,803 (82)	42,371 (71)	6950 (40)	9778 (31)	145,157 (18)	77,840 (13)
Monetary value of all livestock products (% sold to market)	7600 (26)	12,100 (55)	15,030 (3)	13,850 (3)	7500 (0)	7500 (0)
Monetary value of marketable NTFPs (% sold to market)	–	–	250 (28)	1848 (52)	–	–
Total monetary value of marketable produce (% sold to market)	86,403 (77)	54,471 (68)	22,230 (15)	25,476 (18)	152,657 (17)	85,340 (12)
Income from ecotourism	2670	6450	–	–	–	–

pilgrimage centres, but the former was ruled out after inclusion of preferred destinations in the Nanda Devi National Park in the early 1980s and the latter due to new opportunities for income from cash crops after 2000. Shifting agricultural landscapes selected for intensive studies were still too remote and poor in infrastructure to attract tourists.

Pooled information from the intensive studies in the six villages and general exploratory studies in nine villages revealed enormous variation in the spatio-temporal dynamics of biodiversity and driving factors within and between ethnic groups and bioclimatic regions (Table 6.5). Although not observed in intensively studied sites, replacement of potato crop by pea, tomato, *Saussurea costus*, *Inula racemosa*, and hops and conversion of pure crop to a crop-fruit tree mixed production system occurred in other cold desert villages. Conversion of scrublands to

Table 6.5 Changes and driving factors impacting biodiversity and business

Changes	Key driving factors
Cold desert village landscapes	
Partial replacement of potato by pea, tomato, *Saussurea costus*, *Inula racemosa*, and hops	Market demand and financial support from government for all vegetable crops
Apple and apricot planting	Climate change coupled with policy of extending financial support for apple cultivation, value addition, and marketing introduced at lower elevations in the 1960s
Settled agricultural landscapes in central and western Himalaya	
Abandonment of agricultural land use[a]	Increase in crop depredation by wildlife, decline in traditional methods of protection, and uncertainty of full compensation by the government for losses due to wildlife
Conversion of scrub land to agriculture	Freedom of trade of agricultural products by people, restrictions on trade of forest products, and inefficient implementation of forest land protection policy
Decline in cultivation of amaranths[a]	Increased incidence of *Hymenia rickervalis* attack on amaranths
Conversion of agroforestry to agri-horticulture	Policy for supporting horticultural production
Cultivation of medicinal species[a]	Inclusion of naturally rich areas in national parks, increasing demand for herbal medicines, financial support from the government
Cultivation of fodder crops	Decline in nomadic pastoralism, welfare and development programmes tailored largely for settled farmers/landowners/ fodder scarcity
Decline in potato and apple cultivation after 40 years of mono-cropping	Decline in productivity due to pests/pathogens/global warming
Increase in tree cover in farmlands	Risk of losing rights due to absentee landlords/fodder scarcity
Expansion of invasive species[a]	Expansion of road network, abandonment of cropping, and livestock grazing
Demand for *Ophiocordyceps sinensis*[a]	Emergence of new opportunity for income from a medicinal species which had no local uses but the highest economic value among all marketable non-timber forest products
Shifting agricultural landscapes	
Conversion of shifting to settled agriculture/horticulture	Policies favouring settled agriculture, horticulture, and commercial non-forest (rubber, oil palm, coffee, cardamom, and tea) and forest species (bamboo, Toko palm, teak, and *Parkia roxburghii*); non-farm economy and Indigenous innovations in cultivation of broom grass in open fallows and areca nut, betel nut, betel leaf, black pepper, and cardamom under partially opened tree canopy in old growth forests
Conversion of wet paddy systems to fish ponds and cash crops	Quantum of government aid for fishery far greater than paddy cultivation; supply of paddy at subsidized price

(continued)

Table 6.5 (continued)

Changes	Key driving factors
Conversion of area around traditional bridges made from *Ficus elastica* trees/roots into tourist spots and sacred forests as tourist spots	Financial support for development of tourism
In all landscape types	
Formation of village cooperatives attached to companies	Increasing aspirations and financial support for promoting economic growth

Source: observed in rapid surveys and subsequent data collection in the selected landscapes
[a]Changes observed also in the selected landscapes during rapid surveys after data collection

terraced farms, agroforestry to agri-horticulture, cultivation of introduced fodder crops, reversal to traditional staple food crops after intensive potato/apple cultivation for 30–50 years, and increase in tree cover in croplands occurred only in rapidly surveyed settled agricultural landscapes in the moist subtropical/temperate regions. Likewise, conversion of shifting to settled agriculture/plantations, wet paddy systems to fish ponds/cash crop systems, and the transformation of traditional bridges made from *Ficus elastica* trees into tourist spots were observed only in rapidly surveyed shifting agricultural landscapes in the humid subtropical region. We observed no cases of voluntary forest restoration by people, successful forest restoration or procurement of crop/non-timber forest products at minimum support price by government agencies, or CSR activities by business enterprises or farmers' cooperatives/companies in any village. None of the surveyed communities was aware of existing (Green Bonus) and upcoming (UN-REDD) mechanisms for payments for ecosystem services.

Traditional exchanges between local people and nomadic pastoralists or Bhotiya traders no longer exist. None of the villages had marketing cooperatives/producer organizations or access to government agencies procuring farm/forest products at minimum support price. People did not know about electronic/digital marketing systems. All products moved through a local agent in each village to wholesalers in far-off markets. The local agent always delayed payments to farmers though they were paid cash on delivery to the wholesaler. Data collected in the six intensively studied villages showed an increase in product value in the range of 34% (for *Megacarpaea polyandra*) to 210% (*Angelica glauca* and *Carumcarvi*) while moving from farmer to the agent, and 17% (*Nardostachys grandiflora*) to 65% (*Angelica glauca* and *Taxus baccata*) from the agent to the wholesaler. In case of crops, product value increased from 10% (black gram) to 80% (kidney bean) when it moved from farmer to local agent, and 7% (black gram) to 58% (kidney bean) from the agent to the wholesaler. Companies dealing with food/herbal products received commodities from all farmers, local agents, wholesalers, cooperatives, and even retailers, but strictly followed fixed and transparent supply chains for their finished products: through wholesalers/distributors and retailers to consumers. Transactions between farmers and local agents and wholesaler/distributors were not transparent and neither people nor village institutions were fully aware of supply chains and price spreads. People used many species without any market demand for various purposes (21 species as greens, 73 for healthcare, and 5 for fibres).

4 Discussion

4.1 Drawing Evidence-Based Fair Policies

The absence of deforestation and forest degradation and the higher species richness and carbon stocks in the present community forests compared to government forests (Maikhuri et al. 2000; Semwal et al. 2004; Shimrah et al. 2015) testify to the potential of cultural heritage in forest conservation and climate change mitigation (Rao et al. 2022a, b). However, socio-ecological attributes of village landscapes may vary enormously as a result of complex interplay of ecological, social, economic, technical, and policy factors. Species richness and productivity in the present cold desert landscapes are higher than other landscapes with similar biophysical conditions, different ethnic groups, and better accessibility (Saxena et al. 2007, 2011). Ecological degradation around settled agricultural landscapes in Himalaya is an outcome of policies based on assumptions (sustainability of timber-centric forest management) and unethical positions (serving the interests of industries/government over the people who maintained the forest for generations and ignoring detrimental impacts on their livelihoods) (Semwal et al. 2004; Devkota et al. 2017). While unethical positions no longer exist and the pressure of business/industries on forest resources has decreased, uptake of scientific recommendations for restoring the lost ecosystem functions or for preempting future threats is still lacking. Growing crops/fruit trees on degraded forest lands has remained an illicit act since the 1950s, despite research establishing tree-crop mixed farming in degraded forests as a more efficient way of restoring biodiversity, carbon stocks, and economic functions than mere forest tree planting (Bhadauria et al. 2012; Semwal et al. 2013). Nomadic pastoralism and adventure (eco)tourism in Nanda Devi National Park in the vicinity of the settled agricultural landscapes selected in the present study were prohibited in the 1980s and have not been fully restored despite scientific evidence uncovering their roles in biodiversity conservation coupled with conservation of cultural heritage and income for local people (Saxena and Rao 2016; Saxena et al. 2011).

4.2 Enhancing Farmers' Practices

The study shows that local communities have succeeded in maintaining intact forests and agricultural productivity but not in restoring forests, raising agricultural productivity, and satisfying their socio-economic development aspirations.

Participatory research integrates different knowledge systems and stakeholders. Such an approach revealed that (1) a change from traditional-assisted natural regeneration to systematic planting of multipurpose trees and lopping to two-thirds branches; (2) making livestock bedding from broadleaf litter instead of conifer litter; and (3) supplementation of farmyard manure with vermicompost and use of natural pest/pathogen control mechanisms with bio-pesticides made from locally available raw material can reinforce biodiversity conservation and restoration, climate change mitigation, and socio-economic development around settled agricultural landscapes (Semwal et al. 2002; Bhadauria et al. 2014; Maikhuri et al. 2015; Saxena and Rao 2016). Keeping some patches and corridors uncropped can enhance biodiversity and socio-ecological functions of all kinds of agroecosystems (Gayer et al. 2021). Efforts are needed to design such systems in the scenarios prevailing in the Himalaya.

Many farmers in Himalaya gave up the cultivation and consumption of millets under the influence of the mainstream lowland society viewing them as a "coarse poor-man's food" and the supply of only wheat and rice at subsidized prices by government agencies after 1985. The farmers thus lost confidence in the values and uses engrossed in traditional knowledge (Maikhuri et al. 2000). Likewise, they abandoned the cultivation of *Perrilla frutescence* due to the lack of market and demand, although traditional knowledge tells that it protects crops from wildlife (Maikhuri et al. 2000). Despite their high returns for land and prevention of ecological degradation (Maikhuri et al. 2015), intercropping was stopped due to low returns for labour and cropping was abandoned altogether in summer for higher income from collection of wild *Ophiocordyceps sinensis* (caterpillar fungus). Policies on compensating for losses due to wildlife, drought, and flood have also driven decline in climate/wildlife resilient crops and cropping systems and attention to forest ecosystem services that mitigate the losses (Laishram et al. 2020). While people are expected to take a holistic view of social, economic, and ecological implications of abandoning or adopting a practice, government agencies should also be mandated to inform them of the negative sides of introduced interventions. *Ficus elastica* bridges across small rivers and Toko palm roofing are Indigenous knowledge-based parallels to the modern concept of green infrastructure (natural and semi-natural areas designed and managed to meet human needs while enhancing biodiversity and mitigating climate change), but this knowledge has been harnessed for attracting tourists by only a few isolated communities. Promoting community-to-community exchanges thus can also promote the business of ecotourism together with conservation of biodiversity (Fig. 6.3).

4.3 Marketing and Value Addition

A common misconception of people is that businesses/industrial enterprises exploit them due in part to their ignorance of initial investments and market risks. Further, the majority is dejected due to the low economic efficiency of traditional businesses

Fig. 6.3 Intermixing of wild blue sheep and livestock is a scene that attracts tourists in the cold desert. A yak is seen in a herd of blue sheep. (Photo by authors)

and does not realize their strengths in promoting equity and environmental sustainability. This situation has possibly emerged from policies concentrating on delivering material benefits and under-valuing the non-material benefits that underpin Indigenous innovations and thoughtful decisions. The current conventional government market support system is uncertain (e.g. inadequate funds with agencies authorized to procure farm/forest produce at minimum support prices) and inefficient (prevalence of distortions of price spread and coverage largely of mainstream lowlands having low conservation values). Provision of cooperatives framed long ago in the early twentieth century and of self-help groups in the late twentieth century have not succeeded in protecting producers and consumers from unfair practices adopted by intermediaries since ancient times. Technological innovations have led to the electronic National Agriculture Market (e-NAM) in 2016 and the Open Network for Digital Commerce (ONDC) platform in the current year for transparent, free, and fair trade. However, the old evils are still prevalent as most people are unwilling to share the inevitable expenditures in the formation of cooperatives/self-help groups and have a lack of proper infrastructure and expertise for operating electronic/digital systems. Articulations of modern business-to-business, business-to-consumer, and direct-to-consumer models are visible in the traditional knowledge and practices, but people are unable to concretize this knowledge. People tuned to a barter system of business and subsistence economy for generations need to imbibe knowledge on competition and risks in monetary markets, scale of economy, adaptive co-management, and consumer demand dynamics (Berkes 2009; Castillo et al. 2021; Rao et al. 2022b), which is an element lacking in the current policies and programmes. Exposing people to the rare examples of successful transitions to market economies by some communities (Baumann and Singh 2000; Thapa and Rattanasuteerakul 2011) and recognition of the positive dimensions of their cultural heritage (scientific foundation of organic food, herbal medicine, and green infrastructure) could boost morale and trigger efforts towards Indigenous innovations in commercialization of local products and avoidance of exploitation in the market. Effective showcasing of traditional knowledge that is "lesser-known" to national and international communities in formal documents such as the People's Biodiversity Register (Gadgil 2000) can create novel partnerships between local communities, entrepreneurs, innovators, natural healthcare seekers, and interdisciplinary researchers.

5 Conclusions

Interlinkages between biodiversity, business, cultural heritage, and accessibility are apparent in village landscapes, but enormous variation in socio-ecological conditions prohibits generalizations. Improvement in accessibility favours some and restricts other businesses. Policy interventions have ignored the strengths of subsistence-centric traditional knowledge, technologies, and unorganized businesses and the limitations of market/industry-centric scientific knowledge, modern technologies, and organized businesses. Harmonization of people's priorities for livelihood enhancement and global goals for biodiversity conservation and climate change mitigation in a biodiversity hotspot like Himalaya can be achieved by parallel progress in the current states of production, product diversification, value addition, and marketing in village landscapes and by reorienting/expanding traditional cooperation and partnerships in cognizance of contemporary and future scenarios. Participatory action research involving all stakeholders is needed to uncover the science behind traditionally-maintained biodiversity, land uses, and products and to have these results incorporated in the People's Biodiversity Register. Whole-hearted bottom up, objective, and participatory approaches are needed to develop a village landscape-specific package of practices for enhancing the multi-functionality of village landscapes in biodiversity hotspots like Himalaya.

References

Baumann P, Singh B (2000) The Lahaul Potato Society: The Growth of a Commercial Farmers Organization in Himalayan Valleys. Overseas Development Institute, London

Berkes F (2009) Evolution of co-management: Role of knowledge generation, bridging organizations and social learning. J. Environ. Manage. 90:1692–1702

Bhadauria T, Kumar P, Kumar R, Maikhuri RK, Saxena KG (2012) Earthworm populations in a traditional village landscape in Central Himalaya, India. Appl. Soil Ecol. 53:83–93

Bhadauria T, Kumar P, Maikhuri RK, Saxena KG (2014) Effect of application of vermicompost and conventional compost derived from different residues on pea crop productivity and soil faunal diversity in agricultural system in Garhwal Himalaya. Natural Sciences 6:433–446

Castillo JA, Smith-Ramirez C, Claramunt V (2021) Differences in stakeholder perceptions about native forest: implications for developing a restoration program. Restor. Ecol. 29:e13293

Devkota S, Chaudhary RP, Werth S, Scheidegger C (2017) Trade and legislation: consequences for the conservation of lichens in the Nepal Himalaya. Biodivers. Conserv. 26:2491–2505

Gadgil M (2000) People's Biodiversity Registers: Lessons Learnt. Environ. Dev. Sustain. 2:323–332

Gayer C, Berger G, Dieterich M, Galle R, Reidl K, Witty R, Woodcock BA, Batery P (2021) Flowering fields, organic farming and edge habitats promote diversity of plants and arthropods on arable land. J. Appl. Ecol. 58:1155–1166

Laishram J, Saxena KG, Rao KS (2020) Rice cultivar diversity, associated indigenous knowledge and management practices in a lowland village landscape from north-eastern India. Vegetos 33:172–186

Lambooy T, Levashova Y (2011) Opportunities and challenges for private sector entrepreneurship and investment in biodiversity, ecosystem services and nature conservation, Int J Biodivers Sci Ecosyst Serv Manag 7(4):301–318

Maikhuri RK, Rao KS, Nautiyal S, Chandrasekhara K, Gavali R, Saxena, KG (2000) Analysis and resolution of protected area-people conflicts in Nanda Devi Biosphere Reserve, India. Environ Conserv 27:43–53

Maikhuri RK, Rawat LS, Semwal RL, Rao KS, Saxena KG (2015) Organic farming in Uttarakhand Himalaya, India. Int. j. ecol. environ. sci. 41:161–176

Mittermeier RA, Robles-Gil P, Hoffmann M, Pilgrim JD, Brooks TB, Mittermeier CG, Lamoreux JL, Fonseca GAB (2004) Hotspots Revisited: Earth's Biologically Richest and Most Endangered Ecoregions. CEMEX, Mexico City, Mexico, 390 pp

Panwar A, Ober H, Pinkse J (2022) The uncomfortable relationship between business and biodiversity: Advancing research on business strategies for biodiversity protection. Bus Strategy Environ 32:2544–2566

Rao KS, Maikhuri RK, Semwal RL, Saxena KG (2022a) Sacred- the concept in natural resource management in India. Man in India 102:41–64

Rao KS, Semwal RL, Ghoshal S, Makhuri RK, Nautiyal S, Saxena, KG (2022b) Participatory active restoration of communal forests in temperate Himalaya, India. Restor. Ecol. 30(1):e13486

Saxena KG, Liang L, Rerkasem K (Eds.) (2007) Shifting Agriculture in Asia: Implications for Environmental Conservation and Sustainable Livelihood. Bishen Singh Mahendra Pal Singh, Dehradun. 460 pp

Saxena KG, Liang L, Xue X (Eds.) (2011) Global Change, Biodiversity and Livelihoods in Cold Desert Region of Asia. Bishen Singh Mahendra Pal Singh, Dehradun. 322 pp

Saxena KG, Rao KS (Eds.) (2016) Soil Biodiversity: Inventory, Functions and Management. Bishen Singh Mahendra Pal Singh, Dehradun. 462 pp

Secretariat of Convention on Biological Diversity (2020) Global Biodiversity Outlook 5 – Summary for Policy Makers. Montréal

Semwal RL, Maikhuri RK, Rao KS, Singh K, Saxena KG (2002) Crop productivity under differently lopped canopies of multipurpose trees in Central Himalaya, India. Agroforestry Systems 56:57–63

Semwal RL, Nautiyal S, Maikhuri RK, Rao KS, Saxena KG (2013) Growth and carbon stocks of multipurpose tree species plantations in degraded lands in Central Himalaya, India. For. Ecol. Manag. 310:450–459

Semwal RL, Nautiyal S, Sen KK, Rana U, Maikhuri RK, Rao KS, Saxena KG (2004) Patterns and ecological implications of agricultural land use changes: A case study from Central Himalaya, India. Agric Ecosyst Environ 102:81–92

Shimrah T, Rao KS, Saxena KG (2015) The shifting agricultural system (Jhum) and strategies for sustainable agroecosystems in Northeast India. Agroecol. Sustain. Food Syst.39:1154–1171

Thapa GB, Rattanasuteerakul K (2011) Adoption and extent of organic vegetable farming in Mahasarakham province, Thailand. Appl. Geogr. 31:201–209

Wester P, Mishra A, Mukherji A, Shrestha AB (Eds.) (2019) The Hindu Kush Himalaya Assessment: Mountains, Climate Change, Sustainability and People. Springer Nature, Cham, Switzerland, 638 pp

K. G. Saxena Currently an independent researcher in inter-disciplinary and multi-institutional endeavors for the promotion of sustainable farms, forests, landscapes, and livelihoods. Former Dean (Environmental Sciences) and Professor (Ecology and Sustainable Development) of Jawaharlal Nehru University, New Delhi, India.

S. Sreekesh Professor (Geography) at Jawaharlal Nehru University, New Delhi, India, leading programmes on climate change-induced physical and human vulnerabilities and adaptations.

K. S. Rao Senior Professor in the Department of Botany, University of Delhi, India. Currently involved in academic work related to natural resource management and biodiversity assessment in critical ecosystems in hill and mountain regions.

R. K. Maikhuri Professor in the Department of Environmental Sciences, HNB Garhwal University, Srinagar, Uttarakhand, India, with over 38 years of research/development experience in the interphase area of ecology-natural resource management and sustainable development in the Indian Himalayan region.

S. Nautiyal Director, G.B. Pant National Institute of Himalayan Environment (Ministry of Environment, Forests and Climate Change, Government of India), Almora, and Professor, Institute for Social and Economic Change, Bengaluru, India, leading initiatives on natural resource management and socio-economic and ecological approaches to sustainable development.

cc
BY NC SA

Chapter 7
Innovating Products Towards Conservation of the Ifugao Rice Terraces in the Philippines

Jude C. Baggo, Eva Marie Codamon-Dugyon, Clyde B. Pumihic, and Marah Joy A. Nanglegan

Abstract This case study focused on the interconnection of product development and the conservation of the Ifugao Rice Terraces (IRT), a designated World Heritage Site and Globally Important Agricultural Heritage System in the Philippines. It highlights two IRT-based products developed by local meisters, namely, a *tinawon* (*Oryza sativa*) rice-based brew and a taro-based ice cream. Based on cost analysis, the two products showed a higher return on investment (ROI) of 229.4% for the ice cream compared to 118.8% for the rice brew. Both products underwent organoleptic tests and sensory evaluation by consumers and received high acceptance in terms of taste, aroma, colour, and general acceptability. Traditional varieties of *tinawon*, taro, and white beans are used as raw ingredients. The sources of these ingredients range from family produce and local suppliers, such as the Banaue Rice Cooperative.

Key informant interviews and survey questionnaires were used to gather data on the economic potential, sustainability, and experiences of consumers concerning the products. Results showed that the availability of traditional varieties and local suppliers favours the local meisters. This setup decreases their expenses for raw ingredients as compared to purchasing outside their communities, while ensuring the sustainable production of rice brew and ice cream. In the case of the *tinawon* rice, most of the supplies come from the local rice cooperative. The rice supply of this cooperative comes from farmers in the rice terraces. On the other hand, these products added to increasing IRT-based product innovation. In Banaue, as a tourist area, the rice brew and taro bean ice cream are sold to domestic and foreign tourists and consumers. The surveys showed that the fact that the products are perceived as locally produced, organic, and from the terraces adds appeal for the consumers.

Rice brew and taro ice cream are research project outputs under the Ifugao Satoyama Meister Training Programme (ISMTP), a human capacity-building programme initially funded by the Japan International Cooperation Agency (JICA). The ISMTP is a response that contributes to mitigating the enormous pressures faced by the IRT due to climate change, outmigration, ageing population, loss of interest among young people, diminishing Indigenous knowledge, systems and

J. C. Baggo (✉) · E. M. Codamon-Dugyon · C. B. Pumihic · M. J. A. Nanglegan
Ifugao State University, Nayon, Lamut, Ifugao, Philippines

M. Nishi et al. (eds.), *Business and Biodiversity*, Satoyama Initiative Thematic Review, https://doi.org/10.1007/978-981-97-7574-3_7

practices, and other factors. As a programme, ISMTP capacitated 115 IRT stakeholders to become local meisters with research projects geared towards the conservation of thc IRT. Research projects varied depending on the interests of trainees. The overarching principle in the conduct of research projects is the objective of highlighting the interconnection of product innovation and development to the conservation of the IRT. Recognizing these issues and concerns, there is an urgent need to innovate and introduce viable and plausible solutions for safeguarding the rice terraces as an important socio-ecological production landscape (SEPL).

Keywords Ifugao Rice Terraces · Product innovation and development · Taro bean · GIAHS · Local economy and conservation · SEPLS sustainability

1 Introduction

The Ifugao Rice Terraces (IRT) is a remarkable cultural and agricultural landscape in the Province of Ifugao, Northern Philippines. The IRT has been declared a National Treasure, and was designated as a World Heritage Site in 1995 by the United Nations Educational, Scientific and Cultural Organization (UNESCO) and a Globally Important Agricultural Heritage System (GIAHS) in 2011 by the Food and Agriculture Organization of the United Nations (FAO). It is the only GIAHS site in the Philippines.

In this socio-ecological production landscape (SEPL), heritage communities rely heavily on the services provided by the rice terraces for food, medicine, water, wood, tourism revenue, and other purposes. The IRT and its components, such as the forests, rivers, and villages, host the remaining forest cover and biodiversity in the province. Most of the endangered flora and fauna endemic to Ifugao can be found in the IRT such as the *palayon* (*Lithocarpus ovalis*) and *katmon* (*Dillenia philippinensis*) (Taguiling 2013). These trees and plants cover most of the areas in the IRT and serve as a watershed. Indigenous practices in rice farming and natural resource management still persist in the IRT communities. Among these are the *muyung* (woodlot) system of tending forests and the maintenance of irrigation systems through community support actions such as labour exchange.

But amidst this backdrop, the recent Ifugao Rice Terraces Development Master Plan Midterm Review (Rovillos et al. 2022) revealed that the Ifugao Satoyama is suffering from (i) inadequate income of rice terrace farmers, (ii) deterioration of the rice terraces' cultural foundation, and (iii) inadequate support for the conservation of rice terraces. The evaluation noted the decrease in forest cover by 2000 hectares a year in Ifugao, and the introduction of invasive species such as the Asian swamp eel, golden apple snail, and giant earthworm.

A report by Ifugao State University (2014) also noted that the IRT is continuously diminishing in area due to climate change, abandonment, erosion, the introduction of invasive species, the ageing population, and loss of interest among young people. Another report added that the issues faced by this GIAHS in the Philippines include the loss of habitat for agrobiodiversity resulting from land-use conversion, disasters, climate change, agricultural intensification linked to population growth and revenue generation goals, declining interest and capacity among host communities, and inadequate government and public awareness and appreciation of the practical value of agro-biodiversity (DENR-BMB 2013).

During an interview[1] Grace Francisco, a local farmer and ice cream entrepreneur, stated:

> In my community in Ohaj, Banaue, Ifugao, there are problems with the terraces. There is a decrease of indigenous species like edible snails and crops. Also, the climatic patterns change, too. The patterns of rain and drought are getting unpredictable.

On her part, Louiellen Batton, a local meister and business owner said:

> Tourists are getting dismayed when they see destroyed terraces or planted with different crops. Obviously, we in the hospitality business will be affected when the terraces will not be maintained.

Efforts have been made to sustain the multifunctionality of the IRT. Stakeholders and local entrepreneurs have started various capacity development initiatives and product development of non-timber products, but they often are confronted with the issue of sustainability of these business enterprises. The overarching problem is the lack of in-depth appreciation of the interconnectedness of the social, economic, and environmental value of the Ifugao Satoyama as a socio-ecological production landscape. Most of the livelihood programmes implemented by government agencies and other concerned institutions are more anchored on economic purposes and thus miss the important concept of dynamic conservation (Jiao et al. 2022).

1.1 Ifugao Satoyama Meister Training Programme

To address these problems faced by the IRT, the Ifugao State University (IFSU) implemented a human capacity development programme entitled "Ifugao Satoyama Meister Training Programme (ISMTP)," which was initially funded by the Japan International Cooperation Agency (JICA) in 2014 as twinning cooperation between the GIAHS sites in Ifugao, Philippines and Noto, Japan. The programme recognizes that the conservation of the IRT as agricultural heritage should be anchored on strong research projects, product development, policies, and strategic foresight planning. The ISMTP capacitates trainees who are members of local communities after they have qualified for the programme. The ISMTP introduced a trainee-mentor paradigm, in which local trainees are paired with ISMTP mentors composed of experts and practitioners from IFSU, University of the Philippines, University of the Philippines Open University, Benguet State University, Mountain Province State Polytechnic College, and Japanese project managers and partners for research

[1] Conducted on 23 January 2019, at Ifugao State University from 12:00 to 12:30 in the afternoon.

projects and product development. A participatory research approach is also employed in the development of products. In this case, trainees and mentors conduct workshops and consultations on possible research projects to be completed by the trainee. Each trainee presents his/her concepts before a panel of reviewers composed of mentors and representatives from agencies such as the Department of Science and Technology and the Department of Trade and Industry. The panel provides comments and recommendations for the trainee to improve and enhance the research project. Every month from May until February, the trainees present updates on the status of their projects. In February, these outputs are presented during the Philippines-Japan Forum hosted by IFSU. Partners, mentors, trainees, and alumni participate in this forum. To date, the ISMTP has trained 115 locals who graduated from the programme and are now called "meisters."

Enterprising with nature is one of the components of the ISMTP. Under this component, the programme aims to enhance economic opportunities through product development in the rice terrace communities and thereby contribute to the landscape's conservation. As such, it emphasizes the use of local resources in product development. The main elements of the products are either locally grown or found in the Ifugao Satoyama. Commonly used resources include traditional rice and species of wild flora and fauna. Aside from this, the training programme also includes capacity-building activities on the UN Sustainable Development Goals (SDGs), marketing, and packaging. Community exposure is also organized to learn from the direct stakeholders of the IRT about their concerns and issues. Dialogue among agencies, local government units, trainees, and meisters is conducted to highlight the findings of trainees' research projects for policy recommendations.

Since 2014, at least ten products have been developed by maximizing local and available supplies. These products include chips, baby food, rice brew, taro bean ice cream, coffee, rice wine, and traditional and contemporary Ifugao textiles. Other products include gongs and machetes made by Ifugao blacksmiths. These products have become an added livelihood of local meisters while increasing appreciation for GIAHS conservation. Of the products developed, two were commercialized and made it to the market: *tinawon* (*Oryza sativa*) rice brew and taro (*Colocasia esculenta*) ice cream.

Based on the foregoing conditions of the IRT, we examined the interrelationship of product innovation and its impacts on the conservation of the Ifugao Satoyama. Specifically, we aimed to determine the profitability of IRT-based products and their significance in conservation efforts.

2 The Multi-Sectoral Linkages of IRT

The conceptual framework of the ISMTP recognizes product development as a form of intervention in response to the problems in the IRT and as one of the ways to safeguard the Ifugao Satoyama (Fig. 7.1). This conceptual framework places importance on the interdependence of initiatives from a multi-sectoral perspective to

Fig. 7.1 Conceptual framework of the study based on the ISMTP. (Source: ISMTP Orientation Material used by the program created by Judy C. Baggo)

mobilize a community-based set of interventions in biodiversity conservation. These grassroots-based interventions, directly and indirectly, affect the social, cultural, and economic well-being of the communities.

Local businesses that depend on locally-produced raw materials in their production processes. As shown by the two products, their main supplies are derived from the IRT. With this, direct and indirect pressures on local biodiversity, such as deforestation, conversion of the IRT to other land uses, and the changing cultural foundations affect the sustainability of local businesses.

Rice brew entrepreneur, Louiellen Batton, said that

> The supply of heirloom rice comes directly from the producers – the farmers. Even if we like to produce rice-based products but have no supply of rice, this cannot be done. My rice supply is from farmers' rice cooperative Banaue. If they will not continue to maintain the terraces and plant rice, I will not be able to produce these products. The farmers told me that their produce diminished due to various reasons such as broken irrigation, typhoon, and drought.

Figure 7.1 is part of the developed framework being used for the Ifugao Satoyama Meister Training Programme.

3 Project Site

The study was implemented in Banaue, Ifugao, Philippines as shown in Fig. 7.2 (Ifugao 2023). The Municipality of Banaue has a land area of 21,807 hectares comprising 18 barangays and one economic zone with 358 sitios distributed in the vast mountain hinterland. Some of the barangays are traversed by the national road and

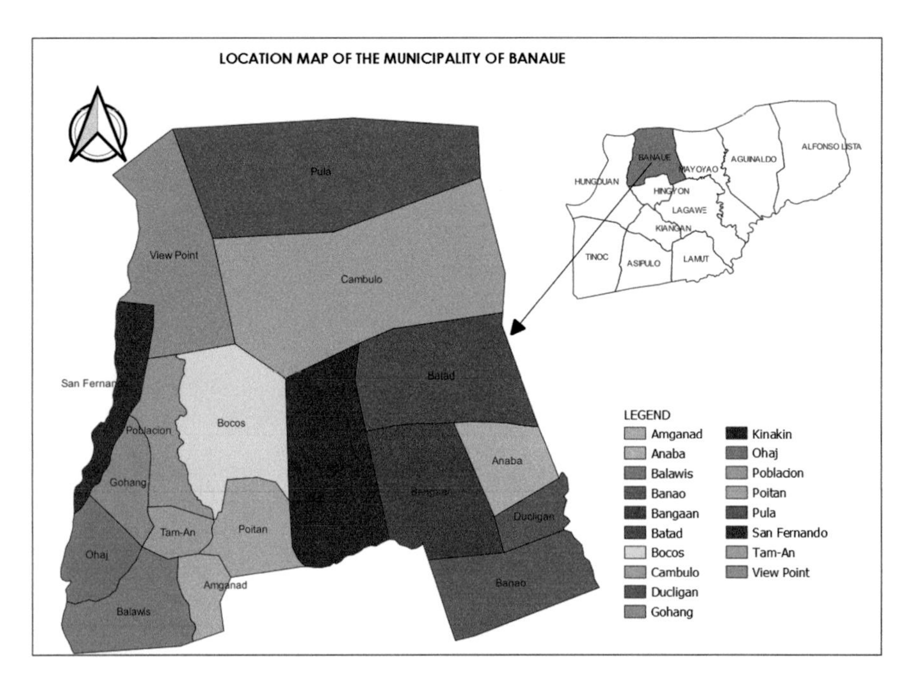

Fig. 7.2 Map of the case study area from the official website of the Ifugao Provincial Local Government Unit (PLGU). The website is state-owned and maintained by the PLGU. (Source https://ifugao.gov.ph/banaue/)

the others can only be reached by foot on mountain trails. The area is classified as a watershed area by the Forest Development Bureau due to its thick vegetation. Of the products studied, the rice brew is made in the town centre of Banaue, while the taro for the ice cream comes from the adjacent village of Ohaj. Table 7.1 reflects basic information about the project area.

4 Methodology

The study utilized a combination of structured questionnaires and semi-structured interviews to determine the profitability of developing products from the rice terraces and examine the interrelationship of these products and the conservation of the Ifugao Satoyama. The instruments were pretested for validation before conducting the actual interviews.

The participating respondents were asked to answer questionnaires during the first phase that included information regarding their profiles, materials, and ingredients from the terraces used in their products, sources of their materials and ingredients, the profitability of their products, and their relevance to the conservation of the rice terraces. For the purposes of this study, two meisters are examined. Louiellen Batton is a local entrepreneur in Banaue, Ifugao. She owns a small hotel and

Table 7.1 Information on the project site

Country	The Philippines
District	Districts II
Municipality	Banaue
Size of geographical area (hectare)	21,807
Dominant ethnicity(ies), if appropriate	Ifugao-Tuwali and Ayangan
Size of case study/project area (hectare)	0.25 hectares
Dominant ethnicity in the project area	Ifugao-Tuwali and Ayangan
Number of direct beneficiaries (people)	
Number of indirect beneficiaries (people)	
Geographic coordinates (latitude, longitude)	16.918610, 121.059174 16.8968, 121.0303 (16° 54' North, 121° 2' East) for Ohaj, Banaue, Ifugao

restaurant. She graduated from the ISMTP in 2019 having completed a study on rice brew (Batton 2019, unpublished). Grace Francisco was a government employee and finished the IMSTP in 2020. Francisco's research project was developing taro bean ice cream (Francisco 2020, unpublished). The second phase of data collection was conducted with walk-in clients at the two business establishments. Eleven locals and tourists were interviewed on their experiences with the products and their views on whether these rice terrace-based products can contribute to the conservation of the terraces. The final phase was a workshop to analyze the data with the meisters and their mentors.

5 Results and Discussion

5.1 *Participants*

Louiellen Batton, a local entrepreneur in Banaue, owns the 7th Haven Café. She is an alumnus of the ISMTP, having belonged to Batch 7 during the 2018–2019 training year. As an owner of a small hotel and restaurant, she incorporated the rice brew into her beverage menu. Customers can choose from the three varieties. On the other hand, Grace Francisco is a former employee of the local government of Banaue. She is also an alumnus of the ISMTP Batch 8 of the 2019–2020 training year.

5.2 *Products*

Tinawon (*Oryza sativa L.*) rice brew, as shown in Fig. 7.3, was developed in 2018–2019 as part of Batton's research project. She produced three variations of rice brew with pure rice, mint, and lemongrass. She used different treatments,

Fig. 7.3 Photos of the rice brew. (Source: Louiellen Batton)

Fig. 7.4 The taro bean ice cream. (Source: Photos by Grace Francisco)

sensory testing, and conducted questionnaires to collect information to improve the taste and packaging. Her supply of rice is from her rice terraces. Aside from her rice fields, she also buys from the Banaue Rice Cooperative. Specifically, she uses the *minaangan* variety of *tinawon* rice for her products.

On the other hand, Francisco's taro (*Colocasia esculenta*) bean ice cream shown in Fig. 7.4 was developed in 2019–2020. Taro is abundant in the terraces of Ohaj in Ifugao. The taro plant is usually intercropped with rice along dikes and stone walls. Francisco also uses dried white bean seeds, white sugar, unsalted butter, coconut juice extract, and salt. These ingredients are purchased in grocery stores.

5.3 *Cost Analysis and Return on Investment (ROI)*

On profitability, both of the products are profitable based on computation of return on investment (ROI) by the following formula:

ROI (%) = (Net Income)/(Total Expenses) × 100 = (Gross Income − Total Expenses)/(Total Expenses) × 100, with an ROI of 0% the break-even point

A. The following is the computation of the ROI for the taro bean ice cream.	
Expenses (in Philippine Pesos):	
2 kg taro	= 70.00/kg
2 glass Nescafe dried white bean seeds	= 40.00
1 kg white sugar	= 50.00
4 tablespoons unsalted butter	= 70.00
2 cups coconut juice extract	= 100.00
Plastic containers	= 140.00
Salt	= 10.00
Labour	= 200.00
Total expenses	= 680.00 PHP (about 13 USD)
Gross income:	
112 containers were sold at 20.00 each = 2240.00 PHP (about 40 USD)	
ROI = (2240 − 680)/680 × 100 = 229.4%	

B. The following is the computation of the ROI for the rice brew.	
Expenses (in Philippine Pesos):	
3 kg *tinawon* (*Oryza sativa L.*) rice	= 270.00
55 rice brew bags	= 32.00
Liquified petroleum gas	= 980.00
Mint/herbs	= 55.00
Labour	= 400.00
Total expenses	= 1737.00 PHP (about 32 USD)
Gross income:	
55 rice brew bags were sold at 70.00/bag = 3800.00 PHP (about 70 USD)	
ROI = (3800 − 1737)/1737 × 100 = 118.8%	

The taro bean ice cream has a higher ROI of 229.4%. On the other hand, the ROI for the rice brew is 118.8%. Both products have different ingredients and procedures.

5.4 *Significance*

Both products have undergone tests and were evaluated through consumer sensory testing by 100 untrained participants from various groups. Sensory parameters were used to determine the acceptability of the products. Three trials were conducted among the treatments. Among the rice brew treatments, the product with lemongrass flavour was the most appealing to participants. In terms of general acceptability, all treatments were rated "liked very much." For the taro bean ice cream, participants found the naturally coloured white beans the most appealing and all treatments had high acceptability.

Both meisters stated during interviews that one of their primary reasons for developing these products was to popularize products from the rice terraces. Francisco said, "Traditionally, my community people only know taro can be cooked and consumed by the household and domestic animals such as chickens and pigs. But with this taro ice cream, people will know that taro can be used in other products that can be sold as a source of income. With this, my people will continue to plant this crop alongside rice." She further noted that the farmers have gained another alternative source of small income derived from their terraces.

Batton highlighted during the interview that, "With my rice brew, I can offer this to local and foreign tourists and clients. But this is not an ordinary brew, it has a story. Aside from introducing our heritage sites, our clients will also appreciate and maybe help in their conservation." She added that the rice roasting procedure is painstaking and takes hours to complete. Batton highlighted the need to further study how to shorten the process.

The second phase of data gathering revealed that all 11 participants expressed their amusement about the two products. One domestic customer suggested that more similar products be developed. He said, "I came to Banaue as a tourist and expected to be offered local products. This rice brew is a good one." Another client stated, "I like local food products. It can be a starter for a long talk and appreciating the culture aside from sightseeing."

Another important observation was the contribution of the products in the local market chain. With the supplies for the rice brew being bought from the Banaue Rice Cooperative, the cooperative gains from the purchase. The funds can be used to buy *tinawon* rice from farmers from the heritage areas. As part of this supply chain, farmers benefit in the process. If this scheme continues, farmers can continue the tradition of maintaining the terraces. The employment of individuals to do the processing of the rice is another positive change in the area.

In addition, both meisters also observed the appeal of locally produced products to consumers. In their interactions with their clients, they said that their customers prefer to buy organic and healthy food products from farmers and establishments. This emerging popularity of organic food drives them to keep producing their products.

Both entrepreneurs along with other meisters participated in trainings on product packaging conducted by partner agencies such as the Department of Trade and Industry. Despite this, Batton and Francisco both stated the need for assistance in the branding and packaging of their products. According to them, aside from the physical packaging techniques, they still need to learn to create more stories about their products.

6 Conclusion

IRT-based product development is viable as shown by the two meisters and their rice brew and ice cream products. The use of available traditional varieties such as rice and taro in product development increases the reliance of local entrepreneurs on these resources from the Ifugao Satoyama. With this setup, stakeholders of the IRT and the business sector reinforce the conservation of these resources. The positive response from consumers proved the growing popularity and profitability of locally developed products. The concept of organic and the sense of helping the IRT stakeholders by buying the products increased the demand for local products. This is a small gain from the ground in terms of linking business and the IRT.

The ISMTP through its mentoring process provided an avenue for local meisters to develop their products. There is still a need to continue innovating other products from the Ifugao Satoyama. This will provide more options for consumers while revitalizing the rice terraces.

On the other hand, both meisters still need training to enhance the branding and packaging of their products. Standardizing the colour, taste, and aroma is still a challenge. A prototype machine to be used in production could simplify the process. Lastly, the best practices of the meisters in developing their products can be replicated in other heritage areas in the province.

References

Batton L (2019) Development of heirloom resources in Banaue, Ifugao. Ifugao State University, unpublished paper

DENR-BMB (2013) Mainstreaming Globally Important Agricultural Heritage Systems into existing policies and programmes. Department of Environment and Natural Resources Biodiversity Management Bureau

Francisco G (2020) The acceptability and profitability of home-made taro-bean ice cream in the Ifugao Rice Terraces. Ifugao State University, unpublished paper

Ifugao (2023) Ifugao Provincial Local Government Unit Website https://ifugao.gov.ph/banaue/

Ifugao State University (2014) Ifugao Satoyama Meister Training Programme Annual Report. Ifugao State University.

Jiao W, Yang X, Min Q (2022) A review of the progress in globally important agricultural heritage systems (GIAHS) monitoring. Sustainability 14(16):9958

Rovillos R, Dacquigan F, Florido R, Baggo, J (2022) Ifugao Rice Terraces Master Plan Review. Ifugao State University.

Taguiling N (2013) Macrofloral Biodiversity Conservation in Ifugao. European Scientific Journal Special edition vol. 4 ISSN: 1857 – 7881 (Print) e - ISSN 1857- 7431

Jude C. Baggo Director for the Globally Important Agricultural Heritage Systems Center of the Ifugao State University. The Center focuses on human capacity building towards the Ifugao Rice Terraces and its attributes.

Eva Marie Codamon-Dugyon President of Ifugao State University. She champions the conservation of heritage, biodiversity, and sustainable development.

Clyde B. Pumihic International Relations Officer of Ifugao State University. His research interests include indigenous cuisines and heritage conservation.

Marah Joy A. Nanglegan Director for the Department of Extension and Training of Ifugao State University. She is an expert in development communication.

Chapter 8
A Culture-Based Social Enterprise That Enhances Soil and Agro-Biodiversity in Guesang, Bangaan, Sagada, Mountain Province, Philippines

Florence Mayocyoc-Daguitan and Guesang Farmers' Organization, Inc.

Abstract This study presents the collective work of the Guesang Farmers' Organization, Incorporated (GFOI) of the Pidlisan Tribe of Sagada, Mountain Province, Philippines, to advance a social enterprise and their right to self-determination. While economic profit remains a central driver for businesses, GFOI stands out by focusing on promoting social values and improving the agricultural lands through increasing biodiversity in the soil.

The discussion draws upon data and information derived through focus group discussions, interviews with key informants, actual observations by the authors spanning from 2014 to the present, and the review of GFOI reports.

The key findings of the study are (1) the setting up of the social enterprise was motivated by GFOI's desire to restore the fertility of the soil by restoring soil biodiversity; (2) restoring soil fertility increased farm yields of farms and encouraged farmers to increase the biodiversity in their agricultural lands; (3) restoring soil biodiversity resulted in the production of a surplus which could be processed and marketed; (4) work arrangements made it possible for workers to still manage their homes and tend their farms; and (5) the enterprise has become a venue for exchange of information, knowledge, and skills, and provides moral support to those in need.

Florence Mayocyoc-Daguitan Coordinator of the Indigenous Peoples and Biodiversity Program at Tebtebba, Indigenous Peoples' International Centre for Policy Research and Education in the Philippines.

Guesang Farmers' Organization, Inc. A multi-sectoral community organization composed mainly of farmers based in Guesang, Bangaan, Sagada, Mountain Province in the Philippines.

Florence Mayocyoc-Daguitan (✉)
Tebtebba, Indigenous Peoples' International Centre for Policy Research and Education, Baguio City, Philippines
e-mail: flor@tebtebba.org

Guesang Farmers' Organization, Inc.
Sagada, Mountain Province, Philippines

M. Nishi et al. (eds.), *Business and Biodiversity*, Satoyama Initiative Thematic Review, https://doi.org/10.1007/978-981-97-7574-3_8

Presently, the social enterprise faces the challenges of meeting a rising demand that exceeds their capacity to supply and the rising cost of fuel and cooking oil, which are not locally available.

Keywords Social enterprise · Soil biodiversity · Agricultural practices · Soil fertility · Traditional knowledge · Motivation

1 Introduction

The Philippines is a country rich in both natural and human resources, yet it remains poor. This phenomenon is deeply rooted in the colonial era's socio-economic institutions which are incompatible with self-determination and democratic governance (Santa Maria 2017). Such institutions enable political dynasties to thrive, who in turn, according to Mendoza et al. (2022), use the wealth of the nation to advance their economic interests at the expense of their constituencies. Despite revolutions, the Filipino people remain poor as corruption and political oppression prevent the country from being lifted up out of poverty (Mourdoukoutas 2017). In 2021, the nation's poverty incidence increased to 18.1%, or about 19.9 million people, from 16.7% in 2018 (World Bank 2023).

Non-governmental organizations (NGOs) and civil society organizations (CSOs) have emerged to fill in the gaps. They have taken on various tasks, such as advocating for economic reforms, civil liberties, and environmental concerns; production and dissemination of information, knowledge, and technology; and strengthening and bridging human capital (Sta Ana III 2002). Some of these made possible the creation of new organizational forms that sought to "pursue a social mission through the use of market mechanisms" (Ebrahim et al. 2014, p. 81, as cited by Habaradas 2022, p. 3). The UNESCAP et al. (2017) reported that social enterprise[1] is vibrant and growing in the country and contributing to reducing poverty and unemployment.

This paper showcases a similar social enterprise of the Guesang Farmers' Organization, Incorporated (GFOI) of the Pidlisan Tribe in the north of the Municipality of Sagada in the Mountain Province of the Philippines. Theirs is an agro-processing enterprise that integrates culture and the Indigenous worldview in the development of a community business. It shall be referred to in this document as the Guesang Social Enterprise.

1.1 Objectives of the Study

This study has two objectives. The first is to illustrate how a business can also serve as a means for social and environmental improvement, in this case a business enterprise that cares for biodiversity conservation. Second, we aim to show how business

[1] https://thegoodstore.ph/blogs/news/what-is-a-social-enterprise

operations can strengthen and sustain Indigenous values. In this case, these include the culture of *ayyew* (no waste) and the culture of sharing, where the emphasis is not just on profit-making, but contributing to the *sumyaan*, or well-being, of the members involved in the business as well as the health of the agricultural production areas.

We have written this study to contribute to abundant evidence that people can do business without having to exploit resources, but rather by caring for the land and advancing people's well-being. It is written with the hope that planners and decision-makers will provide greater support to this kind of development and have the confidence to negate or halt businesses that harm our Mother Earth.

1.2 Research Methodology

Various methods were employed to gather information for this case study. The first and main method was key informant interviews and direct interaction by the authors with members of GFOI during meetings and seminars/training workshops convened by the Indigenous Peoples' International Centre for Policy Research and Education (Tebtebba) and leaders of the Pidlisan Tribe. Secondly, focus group discussions and actual observations were conducted during project visits and GFOI-hosted learning exchange workshops with Tebtebba partners and the rest of the Pidlisan Tribe. The third method was document review, using Tebtebba files and GFOI documents pertinent to the project. Data gathering was carried out from 2014 to August 2023.

1.3 The Study Site

The social enterprise of the Guesang Farmers' Organization, Inc. (GFOI) takes place on the Pidlisan customary land or ancestral domain (Fig. 8.1). This land is located at the coordinates 17°08′N, 120°54′E (Table 8.1). It is bordered in the north by the Province of Abra, in the east by the Municipality of Bontoc, in the south by the Municipality of Sagada, and in the west by the Municipality of Besao. The Pidlisan territory hosts the largest watershed in the Municipality of Sagada in Mountain Province. In 1968 and 1969 with the enactment of Republic Act 3590, the Pidlisan ancestral domain was subdivided into four barangays, the basic political administrative unit of the Philippines, namely, (1) Aguid, (2) Bangaan, (3) Pide, and (4) Fidelisan, the original settlement. Guesang is located on the south-eastern side of Barangay Bangaan.

The people of the land are the I-Pidlisan ("I" means "people of," thus translates to "people of Pidlisan"). They belong to the Kankanaey, one of the eight major ethnolinguistic groups of the Cordillera Administrative Region in Northern Philippines. They are one of the original settler groups in the Municipality of Sagada, who like many Indigenous peoples have strong community solidarity. They are mainly

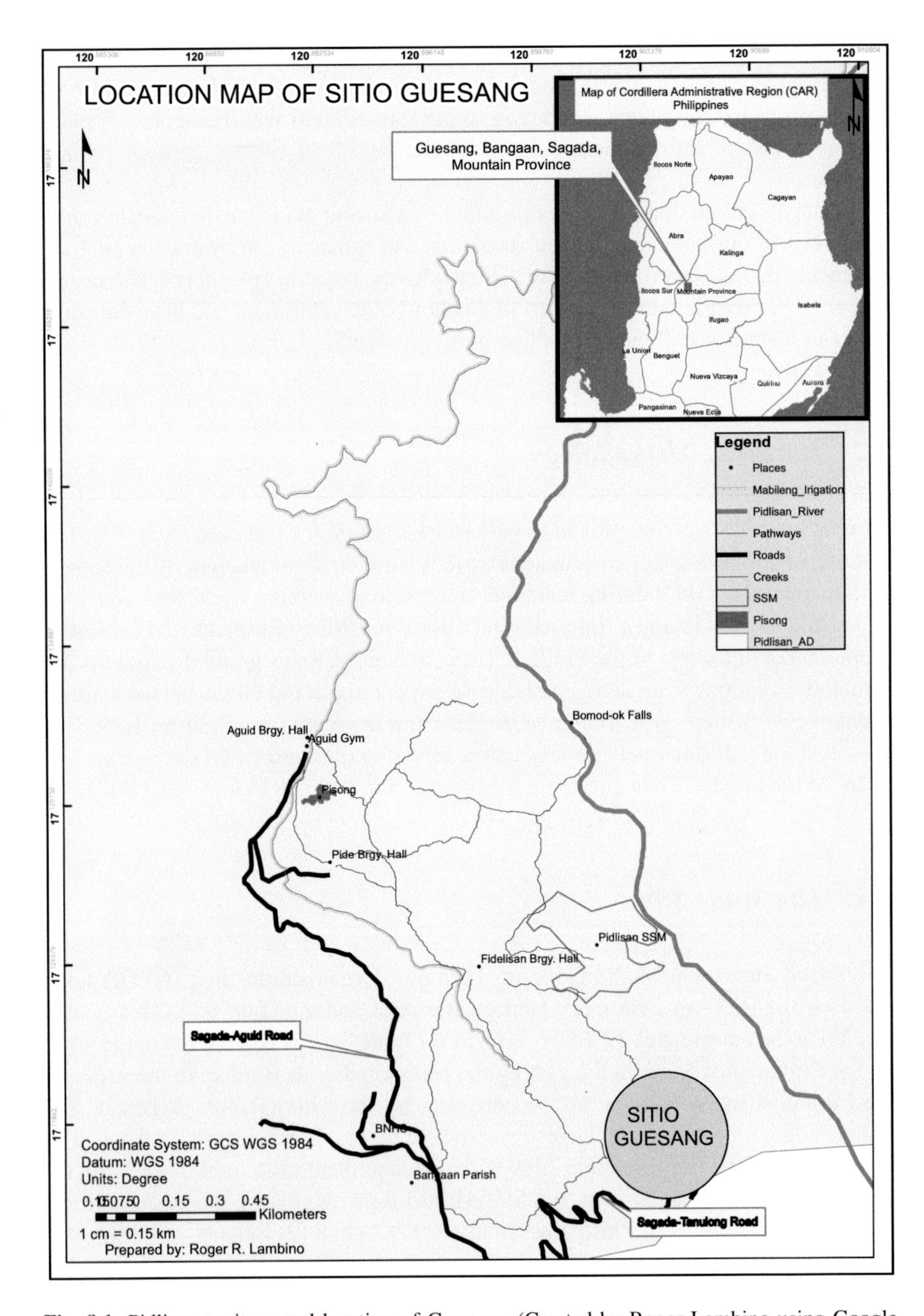

Fig. 8.1 Pidlisan territory and location of Guesang. (Created by Roger Lambino using Google Earth, September 2023)

Table 8.1 Basic information of the study area

Country	Philippines
Region	Cordillera Administrative Region
Province	Mountain Province
Municipality	Sagada
Size of geographical area (hectare)	3902 (community mapping in 2014 with Google Earth 2013)
Dominant ethnicity(ies), if appropriate	Pidlisan-Kankanaey
Size of case study/project area (hectare)	
Dominant ethnicity in the project area	Pidlisan-Kankanaey
Number of direct beneficiaries (people)	82 households; 410 people
Number of indirect beneficiaries (people)	2590 people (Pidlisan) and the buyers
Geographic coordinates (latitude, longitude)	17°08′N, 120°54′E

Table 8.2 Population of Pidlisan by barangay (PSA records)

Barangay	2010	2015	2020
Aguid	596	681	647
Bangaan	703	924	911
Fidelisan	462	428	444
Pide	362	375	423
Total	2123	2408	2425

farmers who have successfully diversified their occupations over time to include weaving, buying and selling businesses, services and enterprises in the tourism industry, food processing, small-scale mining, and wage-earning. Moreover, many are college graduates, and some are employed in government agencies, private businesses, non-governmental organizations, and as overseas contract workers.

Table 8.2 shows that the population of Pidlisan continues to increase. In 2010, the Philippine Statistics Authority (PSA) recorded 2123 persons, followed by 2408 in 2015, and 2425 in 2020 (PSA 2010, 2015, 2020). While their ancestral domain was divided into four barangays, the people are very much aware of their identity as one people belonging to the Pidlisan Tribe.

Traditionally, Pidlisan people organize in various groups, for example, according to where they do their obligations for the common good, like as irrigators, according to their labour exchange network, or according to their geographical location. Moreover, the people have a strong sense of kinship that binds them. This is strengthened by their collective responsibility to conserve and protect the ancestral domain, allowing them to sustain customary land-use patterns and the Indigenous tenurial arrangement, as shown in Table 8.3 (also see Fig. 8.2 for a representation of the Pidlisan landscape).

Table 8.3 Pidlisan land areas by land use and tenurial arrangement (Daguitan and Martinez 2022)

Land use	Area (in hectares)	Percentage	Tenurial arrangement
Payew or rice fields	202.00	5.18%	Family-owned
Batangan or timberland	364.30	9.34%	Tribal and clan-owned forests
Orchard/coffee plantations	16.67	0.43%	Family-owned
Um-a and *uma* in the forests	121.00	3.10%	Family, clan, or tribal lands
Fish ponds	0.24	0.01%	Family-owned
Cemetery	0.73	0.02%	Tribal lands
Residential	29.00	0.74%	Clan/family owned
Granary	1.15	0.03%	Family-owned granaries in the tribal/communal lands
Small-scale mining (SSM)	7.75	0.20%	Tribal/communal lands
Lakes	0.64	0.02%	Tribal and family-owned
Grassland	275.00	7.05%	Tribal/communal lands
Pagpag—pine forest	1214.00	31.11%	Tribal and clan-owned lands
Pagpag (mossy, dipterocarp forest)	1670.00	42.79%	Tribal/communal lands
Total	3902.48	100.00%	

Source: Result of community mapping generated using Google Earth (January 2014)

Fig. 8.2 Pidlisan landscape. (Photo by Guesang Farmers' Organization, Inc.)

Each of these land uses has distinctive implications for biodiversity. For instance, about 30 rice varieties are grown in the irrigated rice fields, and non-aquatic plants such as corn, sweet potatoes, and legumes are planted in the non-irrigated parts of fields. Fields are also a habitat for frogs, mudfish, snails, edible weeds, mole crickets, and insects, some of which contribute to the diet of the people. The *um-a*, or non-irrigated farmlands near homesteads, are planted with the "three sisters"—sweet potatoes, beans, and corn. They also contain fruit trees, such as citrus and bananas. As shown in Table 8.3, the forests are of three types: pine, mossy, and dipterocarp.

2 The Guesang Social Enterprise

2.1 The Birth of the Enterprise

The Guesang Social Enterprise has its beginnings in the partnership of the Tebtebba Indigenous Peoples' International Centre for Policy Research and Advocacy (Tebtebba in short) and the Pidlisan Tribe Organization, Inc. (PITOI) and their common agenda to promote the holistic/integrated, self-determined sustainable development of Indigenous peoples.

The first collaborative work of Tebtebba and PITOI was community action research begun in 2011 to assess the health of Pidlisan territories, resources, and trends in their Indigenous knowledge systems and practices. By 2014, the focus of the research was agricultural land. One significant finding was that the Pidlisan became self-reliant in food after the construction of an irrigation canal in the 1950s, which enabled the people to construct irrigated rice fields, and this self-reliance extended to around the 1980s. It was made possible by the intense cultivation of rice lands and rainfed lands, both rotational farms in the forest (*uma*) and rainfed farms within or near homesteads or rice lands (*u-ma*). People held onto the knowledge and practices of their ancestors and nurtured the land by green manuring, feeding the soil with compost, taking care of forests and communal irrigation canals for sustained irrigation, and bringing in new plants and other varieties of existing plants to increase agro-biodiversity. Green manuring involved the turning over of cuts and pieces of cleared vegetation and selected plants into the soil during land cultivation. Composting, on the other hand, involved piling up cleared vegetation in designated areas of the field.

Another practice is utilizing pig pens as compost pits for all biodegradable waste from the household. These compost materials include the discarded plant parts after harvest, such as rice husks, corn cobs, banana peels, coffee beans pulps, pods from legumes, and plant materials mixed with fruits to hasten their ripening process, among others. Weeds and plant materials within and near the house compound are added, and the stems and leaves of wild sunflowers (*Tithonia diversifolia*) are gathered to be included in the compost pit. These plants with the pigs' manure when composted are hauled into the fields/farms as fertilizers.

Around the 1980s, consumerism was embraced by the people. In addition, attaining higher education became more popular in the community, and tertiary education in the Philippines context is highly commercialized. People had to look for ways to earn cash. Fortunately, they were able to cope with the increased need for cash due to a boom both in the tourism industry and small-scale mining in the community. However, this shift had negative impacts on the farmlands.

Engaging in tourism and mining, where cash is generated much faster, became the preferred occupation of the people. Just before the COVID-19 pandemic, it was common for an I-Pidlisan tourist guide to say, "*I can easily earn enough for one cavan of rice (50 kg) in about four days; whereas to harvest rice on my farm, I would have to work, wait, and maintain the farm for four to six months.*" Accordingly, there has been a continuous decrease in the labour force for farming. As a consequence, the Indigenous farming system has significantly weakened. The cultural practices of maintaining soil fertility by green manuring, composting, and field sanitation can no longer be performed as they require more labour. The main resulting impacts are (1) decreased land productivity, for example, rice harvest decreased by one-third compared to 15 years ago and size of sweet potatoes decreased by half as fusarium wilt infestation reached an epidemic scale; (2) increased presence and lack of control of invasive alien species; (3) abandoned farmlands; and (4) lack of water supply for rice plants. To address decreased yield, people adopted high external input farming (use of synthetic fertilizers and pesticides) and some farms once planted with diverse crops became areas of monocrop production of temperate clime vegetables for the market. Farmers observed that the soil hardened after some years where synthetic/commercial fertilizers were applied, causing them to increase use of commercial fertilizers.

Facilitated by Tebtebba, the Pidlisan community deliberated on the above-mentioned findings. While multiple factors contributed to the decline in agriculture, one prominent factor was identified as the weakening of Indigenous soil fertility management practices, that is, green manuring and composting. Further discussions resulted in a common view that the use of pesticides not only poisons the land, but also contaminates food. People realized that the poisoning of soil, crops, and the environment is in direct contradiction to the principle of *inayan,* which according to Leyaley (2016) means "refraining from harmful acts towards others." *Inayan* is deeply ingrained in the culture of the Pidlisan people, guiding their interactions and decisions. Based on this understanding, they decided to take corrective measures through collective actions to address the increasing use and harmful effects of synthetic fertilizers and pesticides on the soil, other organisms, and humans.

As partners, PITOI and Tebtebba resolved to tackle these challenges by revitalizing and innovating the Indigenous soil management system of the Pidlisan people. The identified innovations included utilization of fermentation techniques to fortify nutrients from plant materials, as well as enhancement of the Indigenous composting process by mechanization and addition of indigenous microorganisms (IMOs). In 2016, Tebtebba provided two shredders for PITOI to mechanize the process of shredding plant materials. PITOI decided to give the management responsibility of one shredder to the Guesang Farmers' Organization, Inc. (GFOI). As follow-up, capacity-building activities were conducted to equip members of PITOI with the necessary knowledge and skills. GFOI hosted training and workshops on producing foliar sprays, culturing microorganisms, and improved composting techniques. One element of the educational module on "How to Capture Beneficial Microorganism and Foliar Fertilizers" is shown in the box below.

Training on Sustainable Agriculture Module II
How to Capture Beneficial Microorganism Foliar Fertilizers

I. **How to Capture Beneficial Indigenous Microorganisms**

A. Indigenous Microorganisms (IMOs)
B. Lactic Acid Bacterial Serum (LABS)

II. **How to Make Foliar Fertilizers**

C. Fermented Plant Juice (FPJ)
D. Fermented Fruit Juice (FFJ)
E. Oriental Herbal Nutrient (OHN)
F. Fish Amino Acid

I.A **How to Capture Beneficial Indigenous Microorganisms (IMOs)**
Materials:

1. 1 kilo half-cooked rice.
2. Any container that can be covered (pail, basin).
3. 1 kilo brown molasses or brown sugar.

Procedure:

1. Cool the half-cooked rice.
2. Put this in container/s and make sure that the container is only about half or two-thirds full to give the microorganisms space.
3. Cover the container/s with the half-cooked rice with a clean cloth or paper.
4. Look for a forest with diverse trees or a bamboo grove.
5. Leave the covered container/s with rice in a place where there are decaying leaves, trees, or bamboo. Be sure that it is protected from rats, animals, rain/water.
6. Collect the container/s after 3–5 days when there is white mould on top. These are IMOs. Remove any black mould as these are bad microorganisms.
7. Transfer the collected IMOs to a bigger container for all collected IMOs.
8. Mix molasses or sugar, same amount as the rice that had been used.
9. Cover the container with cloth or paper and store this in cool place where there is no direct sunlight for 7–10 days.

The product of this is now the pure IMO culture.

Application/How to Use IMOs [General Guide]

1. Mix 2–3 tablespoons of pure IMO into 1 L of water
2. Spray the soil and the plants from 6:00 to 8:00 a.m. or at 4:00 p.m.

Information on how to process this decoction can be found at: https://www.ctahr.hawaii.edu/oc/freepubs/pdf/sa-7.pdf

Note: Translated from Ilokano educational module of Department of Agriculture, a common language that can be understood by the people of the Cordillera Administrative Region

Making use of the shredder, GFOI members pooled their resources and labour to practise what they had learned. They produced organic fertilizers, and captured and cultured local strains of microorganisms that were mixed in to enrich the compost biofertilizers and foliar sprays they produced. They divided these ready-to-use natural/organic fertilizers among themselves, each applying them on their respective farms. Through their sustained efforts, the GFOI members successfully restored the soil biodiversity and fertility of their farms.

Although no formal soil tests were conducted to assess the quality of the soil before and after the application of various organic matter and microorganisms, GFOI members were convinced of the effectiveness of the natural fertilizers based on plant growth performance and the yields of crops. The "soil microorganisms do much more than nourish plants. Just as the microbes in the human body both aid digestion and maintain our immune system, soil microorganisms both digest nutrients and protect plants against pathogens and other threats" (Amaranthus and Allyn 2013).

The efforts resulted in significant improvements in rice lands, for example, successful adopters recorded a yield of 7.2 tons per hectare from a baseline of 4 tons per hectare. Improved growth performance and crop yields were also observed in their *um-a*, non-irrigated farms within homesteads, and their *baangan*, or homegardens. Encouraged by these results, the members of GFOI sustained the cooperative production of fertilizers, maintaining their commitment to the use of locally-produced organic fertilizers, compost, and other materials.

The pooling of labour and resources of the community to address a common need—the organic farm input—was the first business model developed by GFOI. The financial investment was minimal, as almost all materials for the production of biological farm inputs were sourced from within their community. Initially, no monetary gains were realized, but the use of their own products resulted in increased production on their farms. Eventually, some members of GFOI started to produce foliar sprays, selling these to interested people.

The success of these innovations instilled a newfound sense of hope and inspiration within GFOI. Mary Deppas, president of GFOI, expressed this sentiment in

2016, stating, "*Our experience in doing the innovations in the rice lands has inspired us to learn more and to do more and newfound hope was born – that we can do something to better our well-being.*"

2.2 Actors in the Development of the Guesang Social Enterprise

Various groups are involved in the Guesang Social Enterprise. Foremost are GFOI and PITOI as the main actors. Both are traditionally organized but differ in scope of membership. The Pidlisan Tribe Organization was the name given to the tribe by the tribe in the 1990s, and membership is open to all individuals of the Pidlisan Tribe by blood or affinity, as well migrants residing within the ancestral domain who wish to become members. The group clarified its main goal in 2014, which was to strengthen and sustain the traditional unity of the people for the protection and development of their ancestral domain and to safeguard their values and culture. The Guesang Farmers' Organization, as the name implies, is composed of mainly farmers residing in Guesang, a *sitio* in the Barangay Bangaan. The group is self-organized in their practice of *binnadang*, or helping one another.

In line with the promotion of a holistic approach, including a rights-based approach, Tebtebba urged the communities to register their organizations with any authorized government agency, preferably the Security and Exchange Commission (SEC), as this is a requirement to directly access support from both government and non-governmental organizations including lending institutions.

Both organizations simultaneously submitted their registrations with SEC and successfully obtained legal status the same year, gaining the suffix "incorporated." GFOI was in need of more capital for its growing business, and PITOI wanted to access resources for the realization of its goal.

GFOI then took the initiative to establish connections with other civil society organizations. First was with the Episcopal Diocese of Northern Philippines, followed by the Treasure Link Cooperative. They also linked with duty bearers and relevant government agencies, including the Department of Environment and Natural Resources, Department of Science and Technology, Department of Agriculture, and Department of Trade and Industry. These groups and agencies extended different assistance to GFOI, e.g. training on product development and packaging, bookkeeping, construction of the GFOI Processing Centre, enhancement of agroforestry by tree planting, and others. To build GFOI's capacity on organizational matters, Tebtebba sponsored a seminar on leadership and organizational development.

GFOI members have continued to hone their skills to become experts in producing organic/natural fertilizers and are now functioning as qualified resource persons on this subject matter. They were able to spearhead the implementation of a training on organic fertilizer production to the other barangays within the Pidlisan ancestral domain. Their commitment has now extended beyond themselves to upscaling their knowledge, experiences, and techniques to restore biodiversity and enhance soil fertility in the whole of their ancestral domain. This has now become a programme of PITOI with the ultimate aim of eventually getting synthetic fertilizers and toxic pesticides banned within their territory.

2.3 Development and Challenges of the Social Enterprise

GFOI continued their collective effort to sustain and enhance soil fertility, resulting in improved yields on their farms and some surplus produce. Table 8.4 shows the volume of organic fertilizers produced by GFOI through collective efforts for the years 2016 to 2022. Aside from these, some individuals produce their own organic farm inputs.

The continued production and sharing of organic farm sprays has contributed significantly to increased awareness of the importance of a healthy environment in nurturing the health of humans and other animals. Although no tests were conducted to measure the impact of biofertilizers, there are many references to show that microorganisms make a healthy soil and sustain its functionality as a production unit. BASF claims that "It is calculated that one gram of healthy soil can contain up to 1 billion microorganisms and several meters of fungal hyphae [1]. This is particularly important as soil microbes are fundamental to not only soil fertility but also, they are key providers of all types of environmental services" (BASF Agricultural Solutions 2021).

With the increased productivity of their lands, GFOI started processing and marketing their agricultural surplus starting with bananas chips (Fig. 8.3) in 2018, closely followed by muscovado unrefined brown sugar made from sugar cane (*Saccharum officinarum* L.). In 2020, they expanded their product line to include noodles, which quickly became the bestseller among their products (Table 8.5). This expansion into noodle production was made possible through collaboration with the Department of Trade and Industry. GFOI members were trained in the art of mixing various vegetables such as squash, *pechay*, bitter gourd (or *ampalya*), carrots, and root crops with flour to make nutritious flavoured noodles (Fig. 8.4). They are now successfully producing noodles mainly mixed with squash, but also use a combination of in-season vegetables or root crops available locally.

Table 8.4 Recorded volume of organic fertilizers produced collectively by GFOI (GFOI 2022, unpub)

Fertilizer	2016	2017	2018	2019	2020	2021	2022
BIO-SPRAYS (in litres)	50	120	200	220	230	210	190
COMPOST (in sacks)	20	32	50	65	70	65	70

Note: One sack is equivalent to 40–45 kg

Fig. 8.3 Banana chips and vegetable snacks produced by Guesang Farmers' Organization, Inc. (Photo by Guesang Farmers' Organization, Inc.)

Table 8.5 GFOI products and gross sales from 2020 to 2022

	Gross Sales (in PHP)		
Name of Product	2020	2021	2022
Banana chips	89,556.00	113,050.00	224,446.00
Peanut brittle	11,000.00	36,535.00	116,500.00
Vegetable noodles	468,050.00	215,220.00	183,530.00
Vegetable snacks			701,540.00

Source: GFOI 2022, unpub

Based on information gleaned from interviews with key leaders, the growth and expansion of the GFOI enterprise can be attributed to various factors. First is cooperation among members and their commitment to implementing the plans they formulated. Second, they value the knowledge and skills gained from various groups and actively translate insights and learning into action. Third, members are active in participating in project assessments and organizational evaluations, and acting on the recommendations that emerge from these processes. Some examples of how these factors influenced their results are (1) GFOI's efforts to deliberate on and improve their organizational structure towards having clarity on responsibilities, tasks, and authority, and their efforts to improve working relations within their organization, were the outcomes of a leadership skills training; (2) members' initiative to build up their financial management skills by attending seminars and mentoring each other led to the establishment of a Finance Committee that operates with honesty and integrity; and (3) their continued efforts to build linkages with different agencies to access valuable services, such as training on packaging, has led to ensuring product quality.

The organization's success has also been seen in their membership, which has increased over time. Over the past 8–9 years, GFOI membership has grown significantly. They started in 2014 with 14 all-women members and increased to 28 in 2016, 42 in 2018, 63 in 2020, and 85 in 2022 (Fig. 8.5).

Fig. 8.4 Guesang Farmers' Organisation, Inc. processing agricultural produce into noodles. (Photo by Guesang Farmers' Organisation, Inc.)

Number GFOI Members from 2014-2022

90
80
70
60
50
40
30
20
10
0

2014 2016 2018 2020 2022

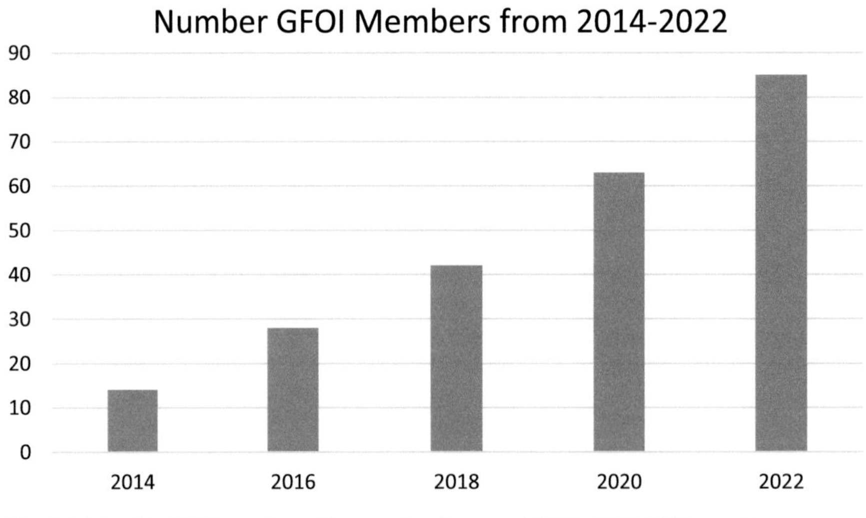

Fig. 8.5 Trend in GFOI members. (Source: As of year-end 2022, GFOI 2022, unpub)

When asked about the key factors that made their enterprise successful, a group of women from GFOI shared their reflections in November 2022. One commented, "*We value whatever assistance is given us, hence we do the best we can to practise or translate into actions the knowledge and insights that we get from training, workshops, and meetings.*"

On the question of the present challenges they face, one woman answered, "*We have to deal with rising costs of fuel and oil, and we are unable to meet the rising demand for our products.*"

The processing of noodles and banana chips, as well as the muscovado production, still depends on external resources such as fuel and oil to run the crusher. Consequently, the rising costs of these external resources has become a challenge for them. Despite this challenge, GFOI members remain committed to their enterprise and continue to process and sell banana chips, muscovado unrefined sugar, and noodles.

2.4 Working Arrangement and Earnings in the Social Enterprise

Around 20 of the women members work full-time in processing centres, with the majority working part-time based on their availability. The working arrangement is flexible, enabling workers to balance their work with other responsibilities, that is, they can still engage in farming, ranging from rice cultivation, farming on non-irrigated farmland, and poultry and livestock raising, as well as manage their families. Thus, they continue to sustain the agro-biodiversity available to the community within and outside farms, manage their homes, and take care of their children.

While only about 30% of members gain income from the processing business, many Pidlisan farmers benefit from the Guesang enterprise as they have a ready market for their farm products, such as bananas, squash, and the variety of vegetables that are mixed into the noodles.

When the enterprise started, processors earned 40 PHP/h, which amounted to 320 PHP/day, or roughly 5.83 USD/day for 8 h of work. By the last quarter of 2022 when GFOI was able to establish its market, the average earning of an individual was about 400–500 PHP/day. This figure was a bit higher than the minimum wage in the Cordillera Administrative Region, which was set at 350–380 PHP/day as of June 2022 (WageIndicator Foundation 2022).

Members emphasized that the benefits from the enterprise went beyond monetary gains. First, the business gave them the opportunity to come together to exchange knowledge and socialize, which helped them in their other work and as the managers of their homes. Second, it increased opportunities to reduce waste by processing surplus farm products. This is important in the context of the Philippines, where perishable farm products go to waste if the price in the market cannot pay for their transport. Third, they acquired skills in food processing, enabling them to provide to other family members and share with relatives and friends, thus allowing them to sustain the practices and inculcate the culture of sharing.

The members of GFOI also emphasized their commitment to keeping their products affordable and accessible to a wide range of consumers, though they have the option of increasing the price of their products. They are keen to increase their production but hesitant to increase their markup price so that these products will be affordable to all and not only to the richer sector of society.

This is despite the fact that GFOI faces rising prices for basic commodities as their enterprise is affected by the volatility of oil and fuel prices.

2.5 Rewards and Recognition

According to GFOI Accomplishment Reports, available from 2015 to 2022, the organization has also been involved in a range of activities that go beyond its core enterprise. The members are involved in reforestation, assisting in the restoration of some degraded portions of the Pidlisan forest. Additionally, they extended credit services to their members and have been involved in infrastructure development, like fixing pathways.

The organization has also availed itself of loans, about two million PHP over time. Members report that these were not solely for their business but covered the family needs of members. Most of these were paid in time according to agreements with the lenders.

Based on interviews with officers (February 2023), it was gathered that the organization has secured several awards and recognitions. In 2018, the organization was recognized as the most performing partner of the Integrated Natural Resources and Environmental Management Project (INREMP) in Mountain Province by the Department of Environment and Natural Resources of Cordillera Administrative Region. On 12 December 2019, GFOI received the best performer award in organizational management by the same agency. On 5 October 2022, the organization was recognized as having the most successful shared service facilities project in the Cordillera Administrative Region by the Department of Trade and Industry.

Currently, GFOI has expanded its operations and work in the six barangays of Northern Sagada, that is, the four barangays in the Pidlisan territory, and Tanulong and Madongo. They have established an internal funding source through membership fees, share capital, annual dues, food processing activities, and income from various projects to sustain their operations and continue their community development initiatives.

Another benefit of the Guesang Social Enterprise is the continuous sharing of knowledge, not only in their business but also skills and knowledge from formal school. As shown in Table 8.6, members' educational attainment ranges from elementary level to college graduate, and some have graduated from vocational schools.

Table 8.6 Number and educational attainment of GFOI members as of the end of August 2023 (GFOI 2023, unpub)

	Total	Female	Male
Elementary level	11	6	5
Elementary graduate	3	2	1
High school level	11	8	3
High school graduate	16	14	2
College level	22	16	6
College graduate	21	15	6
Alternative learning school	2	1	1
Vocational graduate	2	1	1
Unknown	2	2	
Totals	90	65	25

3 Summary and Conclusions

The GFOI Social Enterprise started when people came to realize the bad effects of synthetic fertilizers on the soil and began to make amends. They started their own production of biofertilizers to restore the biodiversity in the soil and to adhere to their Indigenous tenet of ***inayan,*** doing no harm. The social enterprise continued to operate and recently increased the wages of workers at a rate a little more than the minimum wage in the region. Further, GFOI sustained their enterprise as it enabled them to strengthen their practice of ***ayyew***—the concept of no waste or optimizing resources—which is precisely what they practise in their food processing. In seasons of plenty, perishable farm products do not go to waste, but are transformed into other consumable forms or given a longer shelf life. The enterprise is an important opportunity for GFOI to provide other communities with an alternative market for their products and has become a forum for mutual aid among members.

Sustained Indigenous territorial management has been key to the success of the enterprise. Each of the land uses has distinctive implications for biodiversity, thus enabling functioning ecosystems. The farmlands and people are dependent on the forest for water, and people depend on the farmlands for their food and surplus for the social enterprise. Other land uses cater to the tourism industry, while others provide opportunities for wage earning, thus giving community members the capacity to buy local products not produced by their families. All are interconnected.

References

Amaranthus M, Allyn B (2013) Healthy Soil Microbes, Healthy People. The Atlantic, June 2023:3, viewed 15 November 2023. Retrieved from https://www.theatlantic.com/health/archive/2013/06/healthy-soil-microbes-healthy-people/276710/

BASF Agricultural Solutions (2021) Did you know a handful of healthy soil can contain more microorganisms than there are people on earth? Retrieved from https://agriculture.basf.com/global/en/media/public-government-affairs/did-you-know/stories/handful-of-healthy-soil.html

Daguitan FM, Martinez KB (2022) Sustaining our Forests, our Rice Lands, our Culture: Perspectives of the Pidlisan People. Tebtebba Foundation, Baguio City, Philippines ISBN 978-971-0186-39-6

Ebrahim A, Battilana J, Mair J (2014) The governance of social enterprises: mission drift and accountability challenges in hybrid organizations. Research in Organizational Behavior 34:81–100

GFOI (2022) Accomplishment Reports 2015 to 2022, Guesang Farmers' Organization, Inc., Unpublished reports

GFOI (2023) Report on the Educational Attainment of Members as of August 2023

Habaradas RB (2022) Social Entrepreneurship: Conceptual Definition, Brief Literature Review and Some Examples from the Philippines. Department of Management and Organization Ramon V. del Rosario College of Business, De la Salle University, viewed October 10, 2023. Retrieved from https://www.adb.org/sites/default/files/institutional-document/826606/adou2022bn-social-entrepreneurship-definition-philippines.pdf

Leyaley RVG (2016) Inayan: The Tenet for Peace Among Igorots. Int. j. adv. res. manag. soc. sci. 5(2):239–256

Mendoza RU, Yap JK, Mendoza GAS, Jaminola L III, Yu EC (2022) Political dynasties, business, and poverty in the Philippines. Journal of Government and Economics 7:100051

Mourdoukoutas P (2017) Why Filipinos Remain Poor. Forbes, June 2017, viewed on 19 November 2023. Retrieved from https://www.forbes.com/sites/panosmourdoukoutas/2017/06/01/why-filipinos-remain-poor/?sh=4baf43904f9b

Philippine Statistics Authority (PSA) (2010) The 2010 Census of Population and Housing Reveals the Philippine Population at 92.34 Million, viewed 2023. Retrieved from https://psa.gov.ph/content/2010-census-population-and-housing-reveals-philippine-population-9234-million

Philippine Statistics Authority (PSA) (2015) Highlights of the Philippine Population 2015 Census of Population, viewed 2023. Retrieved from https://psa.gov.ph/content/highlights-philippine-population-2015-census-population

Philippine Statistics Authority (PSA) (2020) 2020 Census of Population and Housing (2020 CPH) Population Counts Declared Official by the President, viewed 2023. Retrieved from https://psa.gov.ph/content/2020-census-population-and-housing-2020-cph-population-counts-declared-official-president

Santa Maria R (2017) A Complicated Question: Why is the Philippines Poor? Borgen Magazine, July 2017, viewed 15 November 2023. Retrieved from https://www.borgenmagazine.com/why-is-the-philippines-poor/

Sta Ana FS III (2002) Afterword: NGOs Face Bigger Challenges, Public Policy 6(2):91–107, viewed 29 October 2023. Retrieved from https://cids.up.edu.ph/wp-content/uploads/2022/03/Afterword-NGOs-Face-Bigger-Challenges-vol.6-no.2-July-Dec-2002-3.pdf

UNESCAP, European Union, British Council (2017) Reaching the Farthest First: The State of Social Enterprise in the Philippines, viewed 30 October 2023. Retrieved from https://www.britishcouncil.ph/programmes/society/cso-seed/component-1/reaching-the-farthest-first

WageIndicator (2022) Minimum Wage in the Cordillera Administrative Region, viewed March 2023. Retrieved from https://wageindicator.org/salary/minimum-wage/philippines/2593-car-cordillera-administrative-region

World Bank (2023) Poverty and Equity Briefs, East Asia and Pacific, Philippines, viewed July 2023. Retrieved from https://databankfiles.worldbank.org/public/ddpext_download/poverty/987B9C90-CB9F-4D93-AE8C-750588BF00QA/current/Global_POVEQ_PHL.pdf

BY NC SA

Chapter 9
The Practice of SEPL Based on Weaving-Derived Cultural Business Mechanisms: A Case Study of an Indigenous Settlement in Central Taiwan

Shyh-Huei Hwang, Hsiu-Mei Huang, and Tzu-Hsuan Chan

Abstract This study takes the Zhongyuan settlement of the Seediq in central Taiwan as a case in point to explore from a small spatial span the relationship between Seediq traditional weaving, which is closely related to the natural environment, and SEPL in the development of cultural business mechanisms. The settlement is proud to have a large number of people with traditional weaving skills and strives to promote traditional weaving and innovation. Under the pressure of cultural and commercial mechanisms, it still hopes to slowly progress towards the dream of becoming a weaving craft village. The preliminary conclusions of this study are (1) from the orientation of the relationship between weaving skills and the environment, the case embodies the wisdom of human beings to respond to the natural environment; (2) from the orientation of the environment and landscape, homestead ramie fields have been restored but their function has changed from meeting daily needs to teaching or cultural experiences; (3) traditional weaving is limited by slow manual speed, high labour costs, and the mission of cultural preservation, so profits are limited. Adjustments to methods geared to developing financial resources for diversified development include winning funding support from government departments based on cultural preservation and satoyama contributions, and enriching ecological and environmental knowledge for cultural tours as a cultural business mechanism; (4) cooperation between the settlement and the university can play a small role in the content of the proposals on the satoyama spirit; (5) revitalization of weaving and tribal guiding are two cultural business mechanisms that form the protective umbrella of the Zhongyuan settlement SEPL.

S.-H. Hwang · H.-M. Huang (✉)
Graduate School of Design, National Yunlin University of Science and Technology, Douliou, Yunlin, Taiwan

T.-H. Chan
Graduate Institute of Architecture and Culture Heritage, Taipei National University of the Arts, Beitou Dist., Taipei City, Taiwan

M. Nishi et al. (eds.), *Business and Biodiversity*, Satoyama Initiative Thematic Review, https://doi.org/10.1007/978-981-97-7574-3_9

Keywords Cultural business mechanisms · Traditional weaving of indigenous people · Homestead of ramie field · Seediq tribe · SEPL umbrella

1 Background

The population of Taiwan is 23.45 million, and the Indigenous population of Austronesian peoples is about 580,000, accounting for about 2.47% of the total population. There are currently 16 officially recognized Indigenous ethnic groups, each with their own unique culture that enriches Taiwan's multicultural society. Compared with the Han people who make up the majority of the population but have not engaged in weaving for a very long time, many Indigenous people traditionally weave and wear the woven clothes themselves. The main material is ramie, which plays an important role in the lives of the people. In addition to being used as a weaving material, ramie can also be made into hunting backpacks and ropes for binding. Unlike those in larger-scale agricultural landscapes such as terraced fields, the Indigenous people in Taiwan typically plant small ramie fields near their homes for their own use. Accordingly, small landscapes formed within their settlements. From the cultivation and harvesting of ramie to the weaving of ramie yarn, processes are complicated. Likewise, the dyeing of the ramie fabric is both performed in the environment and closely related to the environment. In addition to the learning and presentation of skills, weaving is also affected by the norms of ancestral beliefs, social interaction, and other related factors.

The Seediq are one of the 16 ethnic groups in Taiwan and are famous for being good at weaving. In the past, almost all women could weave and had the above-mentioned knowledge and abilities related to weaving. However, modern clothes worn by the tribe today have replaced the clothes they once weaved themselves. The lack of weaving in daily life has gradually led the tribe to stop weaving. Traditional skills are endangered, and ramie fields have been abandoned, resulting in the weakening of the original cultural ecosystem based on weaving. Further, the relationship between the tribespeople and nature has gradually become more alienated, weakening the **socio-ecological production landscape** (SEPL). In recent years, due to the revival of Indigenous culture, Indigenous ethnic groups have begun to develop cultural industries suited to each group's characteristics. A few Seediq women have gradually resumed weaving and planting ramie. Weaving and the cultivation of ramie fields have also changed, moving from being internal to ethnic women's homes to the small landscape of external cultural representation and reproduction.

This case study examines the Seediq tribe in the Zhongyuan settlement of Ren'ai Township. From a small spatial span, we explore the role of Seediq traditional weaving, which is closely related to the natural environment, and SEPL in the development of cultural business mechanisms. The study focuses on the actions of the Weaver's Home (WH), the only entity among the settlements that uses weaving as a culture-based industry (and arguably the most active representative of the ethnic group). In the process of diversified development in the present day, the cultural business mechanism of weaving is used as a protective umbrella for the maintenance of the SEPL. The history and types of actions by the various actors, as well as their limitations, are analysed. Ultimately, the vision of a weaving skill settlement and the possibility of harmonious coexistence between humans and nature (at the core of the spirit of the satoyama) are explored.

1.1 Seediq, Known for Their Weaving

The Seediq population of 10,908 is composed of three groups—Tgdaya, Toda, and Truku, mainly distributed in central Taiwan. Ren'ai Township in Nantou County is the ancestral home and has the largest population (Iwan 2014). The early lifestyle of the Seediq was based on the shifting cultivation of mountain fields and hunting to obtain food, that is, they were a people who depended on the mountains and forests for their livelihoods (Tian 2001).

The Seediq people believe in the immortality of the soul and refer to their ancestors' instructions as their *Gaya* beliefs.[1] In addition to regulating moral standards and ritual taboos in daily life, *Gaya* also regulates the division of labour between men who hunt and women who weave cloth. Traditionally, men had to hunt to become warriors, while women had to learn to weave cloth before they could have their faces tattooed, and only after they had done so could they be married (Wang 2003; Yeh 2011). Only brave men and women good at weaving are allowed to cross the rainbow bridge after death and join their ancestors in glory. These beliefs resulted in men being familiar with the knowledge system of the mountains and hunting, while women were familiar with knowledge related to farming and weaving, and took weaving as their vocation (Tseng 2013; Hwang and Huang 2019).

1.2 Space and Ecosystem of Zhongyuan Settlement

This study takes the Zhongyuan settlement located in central Taiwan, in the northwest corner of Ren'ai Township, Nantou County, as an example. The Zhongyuan settlement is affiliated with Huzhu Village, with a settlement population of about 800 people (Fig. 9.1a, b and Table 9.1). The tribe belongs to the Tgdaya language group of Seediq. The tribe's original place of residence is called the Balan Community. They once held important hereditary positions within the Seediq (e.g. priests). Due to their high social status, they uphold high standards for themselves

[1] The *Gaya* regulates the moral standards and ritual taboos of the tribe. Those who abide by the *Gaya* will be blessed by the ancestral spirits (*utux*), while those who do not will be punished. This is why the *utux* are treated with absolute obedience, and children are taught from an early age that they must be careful not to offend and that only by fully observing the *Gaya* can they receive the blessings of the ancestral spirits.

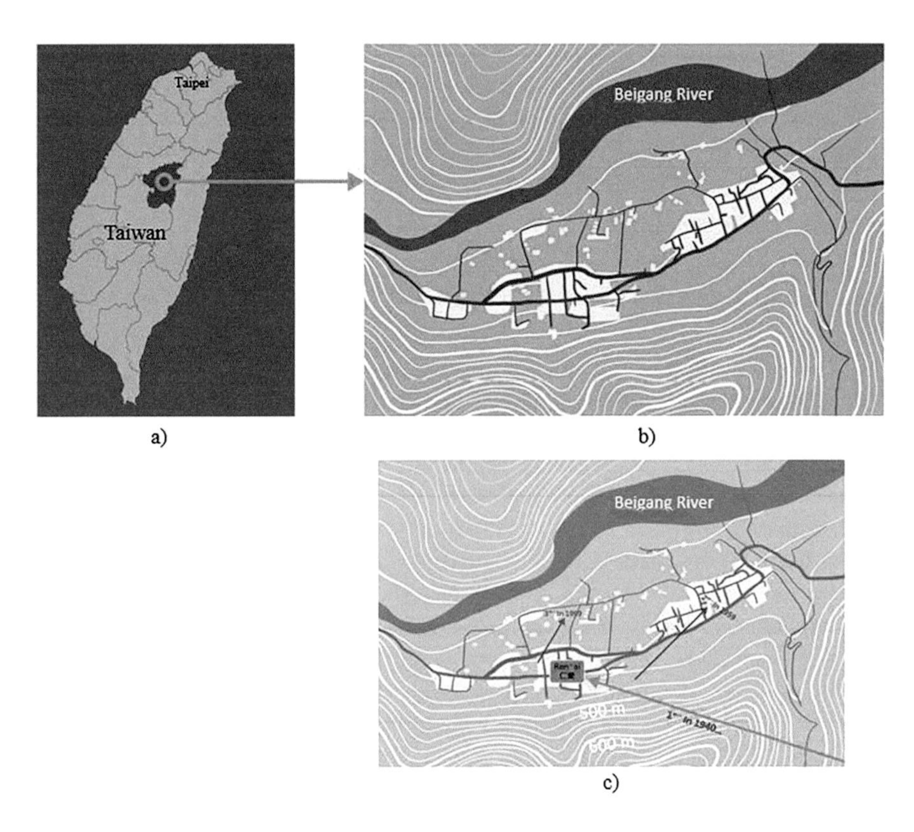

Fig. 9.1 Zhongyuan settlement location and landscape. (**a**) Location, (**b**) Satellite map of Zhongyuan Settlement, Ren'ai Township, (**c**) The migration of tribes in the past hundred years (in 1940, the whole village moved from 40 km away. In 1959, due to floods, in 1999, due to earthquake, the villagers moved within the village), (**d**) Landscape of the Zhongyuan Settlement SEPLS, Ren'ai Township, Nantou County, Taiwan. (Source: (**a**), (**b**), and (**c**) base map from ©Google (2023) with content drawn by the authors, (**d**) Photo taken by authors on 9 September 2020)

in the weaving skills that the tribe values. However, during the Japanese rule in 1940, the people of the settlement were forced to migrate to the Zhongyuan settlement about 40 km away from their ancestral home due to the construction of a reservoir (Shen 2008). The altitude where they lived dropped from the forest area at about 1200 m above sea level (m.a.s.l.) to a river level of 500 m.a.s.l. On the terraces, the tribespeople's livelihoods range from hunting to farming. Over the last eight decades, this tribal group has faced the fate of being forced to migrate many times. In addition to the above-mentioned relocation of the entire village in 1940, they also experienced small-scale migration within the village due to floods in 1959 and an earthquake in 1999 (Fig. 9.1c).

Table 9.1 Basic information of the study area

Country	China
Province	Taiwan
District	Nantou County
Municipality	Ren'ai Township (Huzhu Village)
Size of geographical area (hectare)	3100
Dominant ethnicity(ies), if appropriate	Indigenous people of Seediq
Size of case study/project area (hectare)	1500
Dominant ethnicity in the project area	Indigenous people of Seediq
Number of direct beneficiaries (people)	700
Number of indirect beneficiaries (people)	10,000
Geographic coordinates (latitude, longitude)	24°03'56.7"N 120°58'00.6"E

Source: prepared by authors

Surrounded by mountains, the Zhongyuan settlement is divided into the Ficus-Machilus zone and the Machilus-Castanopsis zone[2] according to the vertical vegetation zones of Taiwan's mountains (Liu and Su 1983). Both are low-medium altitude broad-leaved forests with high species diversity and rich ecological systems. A survey conducted by Liao et al. (2011) recorded a total of 438 plant species and 122 plant uses by community members interviewed. Among plants used, 40 species related to hunting were the most common, followed by 38 species of edible plants. Others were used for utensils, agricultural plants, building materials, medicine, clothing, and ritual plants. For example, the Taiwan euonymus, the material of which is elastic and not easily deformed, is the main plant used by hunters to make bows. The Taiwan crepe myrtle is a good fuelwood, and its leaves can be used as dye for weaving. The first author of this study led students to interview tribal people from 2021 to 2023, and investigated about 60 species of plants. The students also investigated and found that, for example, traditionally, hunting is men's business and usually requires living in the mountains for many days. The tribesmen will use the leaves of the Taiwan sugar palm to make the roofs of temporary huts. Ramie is not only used to weave clothes, but can also be woven into hunting belts. Weaving utensils for living things is also a matter for men. They go to the mountains to collect Yellow Rotang Palm and weave them into backpacks (Fig. 9.2). They also use the beautiful Galangal to weave into small baskets and mats (Fig. 9.3).

The settlement is located on the river terrace of the Beigang River. The soil is fertile. According to a 2007 survey by the Nantou County Government, the land

[2] In Taiwan, landforms range from flatlands to mountains higher than 3500 m.a.s.l. The distribution of mountainous vegetation zones depends on altitude, latitude, and temperature, while the climatic zones range from tropical to sub-frigid zones. For the area described in this chapter, the altitude of the Ficus-Machilus zone is <500 m.a.s.l. and is a tropical climate, while the altitude of the Machilus-Castanopsis zone is 500–1500 m.a.s.l and is a subtropical zone. Source: https://kids.coa.gov.tw/view.php?func=knowledge&subfunc=kids_knowledge&category=B11&id=27&print=1

Fig. 9.2 Backpack made from Yellow Rotang Palm collected on the mountain. (Photo taken by authors on 9 September 2020)

Fig. 9.3 Small baskets and mats weaved using the stems of the beautiful Galangal. (Photo taken by authors on 6 January 2023)

area of Ren'ai Township is 127,353 hectares, and the cultivated land area is 7433 hectares. Most farmland is used for dry field crops, with crops such as tea, fruits (including plums), and vegetables accounting for 93.5%. Paddy fields make up only 6.5%, and most are distributed in the Zhongyuan settlement at a lower altitude in the township and the neighbouring Qingliu settlement (Nantou County Government 2007). In the 4 years that our team has been observing and interviewing tribespeople, the situation has remained generally the same. However, in the early years there were more rice fields, and in recent years more and more vegetables and fruits are grown. Plums are an important economic crop for the settlement. For sightseeing purposes, local residents organize flower-viewing activities in January when the plum blossoms are in full bloom, as well as plum-picking experiences in April.

1.3 Seediq Weaving Knowledge System with Strong Environmental and Social Ties

Traditional crafts are part of the traditional knowledge of human communities and a channel for cultural expression. They are practical and technological traditions developed by local communities using local materials in response to their natural environment and social contexts that are carried out to meet their living needs. The local knowledge of communities also serves the function of maintaining "locality" and local identity (Chiang 2016).

In the early days of mankind, the use of natural materials for daily needs was the norm. Taiwan's Indigenous crafts are based on connections to living areas, beliefs, religions, social classes, and clan organizations (Wang 2001). Since traditionally there is no system of writing, oral transmission, visual symbols, and rituals have long been an important means of transmitting information and cultural inheritance among the settlements. The skills of the Indigenous people are deeply rooted in the traditional cultural lineage of the ethnic group, reflect group consciousness and land and regional characteristics, are closely integrated with daily life, and focus on applied visual art expression (Wan 2009). The knowledge systems involved in Indigenous crafts are usually holistic, being practical knowledge developed in the face of both the world of existence (including society and nature) and the philosophical foundations of a long-established worldview. The various fields are not fragmented; however, they interact with each other and influence each other. If we only think in terms of modern subject categories, we will only see fragments and fail to truly grasp the content of Indigenous culture and knowledge (Chen 2009).

The weaving by the Indigenous Seediq people of Taiwan, based on the team's interviews with the settlement, is estimated to have been active until approximately 40 years ago. Then, it was still customary for community members to grow their own ramie and weave it into yarn after processing for their family's clothing and quilts, thus entailing a holistic environmental and social knowledge system. Although in recent years people have replaced weaving with buying ready-made clothes, the knowledge system still exists in the memory of the older generation.

1.3.1 Ramie Becomes Woven Yarn After a Complex Treatment Process

Seediq women weave cloth with yarn made of ramie, which they usually grow for their own use. A complex process is involved, from planting ramie to making ramie yarn (Fig. 9.4). Ramie planting does not require special care. It takes about 3–4 months from planting to harvest as long as there is sufficient sunlight and water.

Fig. 9.4 Turning ramie into yarn requires a complex process of more than ten steps. It takes about 1 month of work to make 1 kg of ramie yarn that can be used for weaving. (Photos taken by authors in 2020)

Ramie can be harvested when it reaches a height of about 150 cm, usually three times a year. After a year's harvest, the roots of ramie are dug out, the field is turned over, and then replanted. Growing ramie is a woman's job and the ramie needs to be scraped immediately after picking. Busy women often estimate acreage for planting based on their own time and labour. Therefore, the average family can only plant about 50 m^2 of ramie near their home. Due to the complex production process, a ramie field of this scale can only produce about 10 kg of yarn a year. Therefore, unlike large-scale agricultural landscapes such as terraces that are used for economic income, Seediq ramie fields are small in scale and planted by women in small quantities near their homes for their own use. The ramie field is considered part of the homestead.

Processing ramie yarn is time-consuming. It is customary for women to hang yarn around their necks so that it can be joined at any time, even during the odd hours of walking in the fields. The joined yarns are put into a basket with harvested vegetables. After joining and twisting, the yarn is typically used for two purposes. The uncooked thread is called raw ramie and can be used directly as rope for binding, or woven into net bags for hunting prey (such as wild boar and muntjac deer called mountain *Qiang*). The second use involves more processing, where the ramie is boiled and dyed using natural materials from around the settlement to be made into woven cloth for clothing. First, the yarn is boiled with wood ash and after a few hours, it softens and turns white. It is then called boiled ramie. Traditional Seediq garments are white, red, and black. White yarn is made by boiling the ramie yarn with the ashes of *Trema orientalis* (L.), red is made by boiling and dyeing the white yarn with Shoulang yam several times, and black yarn is made by soaking white boiled ramie with black mud and pounded Taiwan crepe myrtle leaves.

Yarn processing is a daily task for women and therefore, in addition to the technical skills required, it also extends to many social norms. For example, if you find a ball of yarn dropped on the road, you must not keep it for yourself, but hang it on a branch so that the owner can easily retrieve it. Pregnant women are not allowed to spin or twist yarn, otherwise the baby's umbilical cord will be wrapped around its neck at birth.

1.3.2 Weaving Skills as a Complete Cultural Ecology

The Seediq weaving is extremely well known among the Indigenous peoples of Taiwan. Traditionally, women weave cloth for clothes and quilts for their families. Weaving tools are simple and made from natural sources, consisting mainly of warp frames, warp boxes, and other parts of the tools. The making of the tools is a man's job, sourcing wood and bamboo from the nearby mountains. The largest and heaviest warp boxes are usually made from harder woods such as beech and Formosan Michelia (*Michelia formosana*). Before beginning the weaving, the warp process must first be finished. Therefore, the people of the settlement must put aside their chores and children to concentrate on finishing the warp needed for a piece of cloth. If something goes wrong, the yarns need to be completely dismantled and refinished. It usually takes two people working together to remove the yarn from the warping frame and put it on the loom.

To begin weaving, the woman sits on the floor and uses her back strap to hold the warp rolls together with her body, giving her an L-shape. She then stretches her legs flat against the loom (weaving box). In weaving, the tension between the waist and the feet is coordinated, and the body moves in rhythm as the weft threads are woven in between the tightening and the loosening of the weaving box with the feet, as the pattern of the woven fabric is fabricated. The traditional loom is a simple tool that allows for the development of many different techniques and patterns, but it also tests a woman's physical strength and stamina as she cannot move freely (Fig. 9.5).

In addition to the needs of life and the influence of *Gaya* mentioned above, weaving also has a social function and significance. For example, the skill is traditionally passed on from mother to daughter. Mothers would weave cloth for their daughters as a dowry for the future, and elderly women would also give their own cloth as a gift to add to the dowry, so that the cloth brought by the bride could be given to her husband's relatives as the first bridge to build interpersonal relationships.[3] It is said that the more cloth in the dowry, the more attention the bride will receive from her relatives and friends, and the higher her status will be in the family in the future. In addition, according to the elderly, a girl's parents can ask the boy's family to give

[3] Seta Bakan, a woman interviewed by the team and member of the settlement whose grandmother and herself are both good at weaving, had a 100 pieces of cloth for her dowry when she got married around 1980. The team interviewed about 20–30 women of the settlement, of whom all women over 45 years of age still had cloth for their dowry when they got married.

Fig. 9.5 Seediq traditional weaving. (Photo taken by authors on 12 December 2021)

more betrothal gifts when proposing marriage in the settlement if their daughter is good at weaving.[4]

The Seediq have a customary division of labour between men and women, with the men hunting and defending their homes while the work of farming and weaving is performed by the women. The people place a high value on a woman's ability to weave. Women who cannot weave are called *ngangah* (meaning fool) or *uxe mqetin* (not a woman) (Cao 2021). As weaving involves many complicated processes and women are busy with farming during the day, weaving work is carried out in early morning and at night in the women's spare time. In short, Seediq women are extremely busy and hardworking. Even so, these factors still drive them to consider weaving as their vocation.

From the above descriptions of the knowledge system of weaving, it can be observed that the traditional knowledge system of weaving encompasses natural knowledge (e.g. geographical, environmental, local materials for tools, ramie planting technology), *Gaya* beliefs, generational links, and social interaction. Traditional weaving presents an example of the wisdom of humans in response to their natural environment (Fig. 9.6).

[4]The betrothal gift is usually based on the number of pigs, but if the daughter is a good weaver, the parents may ask for more pigs.

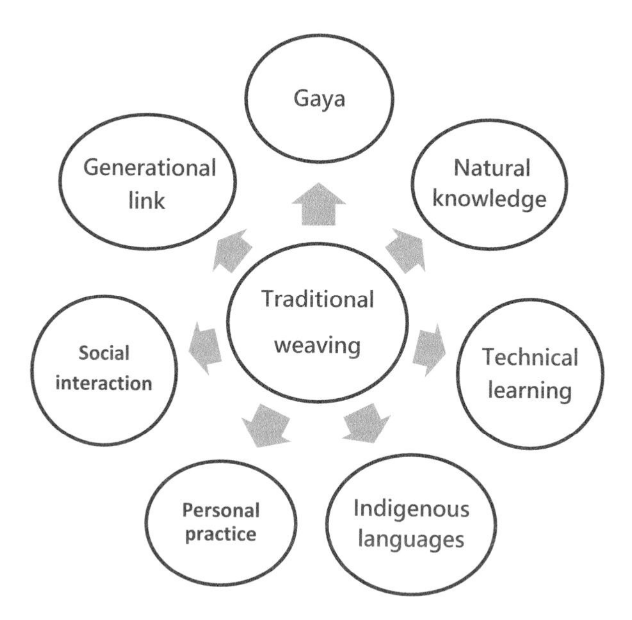

Fig. 9.6 Seediq traditional weaving skills as a complete cultural ecology. (Source: prepared by authors)

1.4 Changes in the Weaving Landscape: A Threat to Traditional Skills

1.4.1 No Need to Weave Clothing in Present Day

Nowadays, due to changes in the social structure, the settlement people do not need to weave their families' clothing, and the *Gaya* norms are being relaxed. As a result, weaving skills are on the verge of being lost, leading to increased alienation from the environment and the disruption of the knowledge ecosystem, such as the lack of need to grow and handle ramie, the loss of opportunities to learn about the environment, and the weakening of social network connections.

1.4.2 Blind Spots in the Promotion of Cultural Commoditization

Starting around the 1990s, Indigenous communities actively developed cultural industries and encouraged their people to open shops for weaving products. Seediq weaving was also developed and flourished in this way (Wang and Iwan 2012; Ho 2010). However, the focus was on commodity sales based on the concept of tourism focused on different ethnic groups. Cultural tourism became a form of consumption, a secular mass-produced commodity that was branded and presented on stage as exotic (Goss 2005). Scholars have also expressed concern that the culture of Indigenous communities (e.g. weaving, carving, dance) has been modified to suit the preferences of tourists, which may be detrimental to the settlements (Hu 1993; Hsieh 2004).

The commoditization of Indigenous crafts was so popular in the 1990s that market logic led to price competition. A few shops even used fake factory-made cloth or foreign goods made with cheaper labour. This trend was not helpful for the preservation of skills and culture in settlements. In recent years, tourism has developed in a more diversified way. For example, tourist experiences can help outsiders understand ethnic characteristics, but only give them a superficial taste of the technology. If the traditional loom is replaced with one introduced from a foreign country that is easy to operate, a blind spot remains. Such experiences cannot reflect the overall knowledge system of traditional weaving, and present only fragmented knowledge. In the Seediq settlement, a small amount of ramie may still be planted, but it may only be used as a demonstration for tourists and not really made into woven yarn. Therefore, developments are turning in a direction whereby internal and external collaborators in the settlement work together to re-examine and reorganize the relationship between the weaving system and the environment.

2 Research Method and Actions

The methodology of this study is based on qualitative research methods using participant observation and interviews. For participant observation, from 2014 to 2017, the second author of this study participated in a course on weaving techniques offered by cultural asset preserver Seta Bakan. The course focused mostly on weaving techniques, while ramie yarn making was more experiential in nature. From 2020, the research team began visiting Seta Bakan's hometown, the Zhongyuan settlement, to conduct interviews with weaving practitioners. The follow-up group of authors was then divided into two parts. The first continues in-depth interviews on the people's views on weaving skills and ideas on settlement development, conducting research and analysis. The other part has worked with the settlement from 2020 to collaborate on university courses on "Community Reconstruction" and "Cultural Industry" offered at the first author's university. Postgraduate students in design are brought to the settlement to conduct workshops every 6 months. To date, six workshops have been held and they are ongoing (details on workshop topics are discussed in Sect. 3.2.2). The research team also uses an action research method to review and improve each workshop and consult on the tribe's needs to construct an action strategy for the theme and content of the next workshop. The team serves as both event planner and participant observer. It observes students as they interact with tribespeople, reviews and modifies the results submitted by students, and then gives these back to the settlement.

3 Analysis and Discussion

This study takes the Zhongyuan settlement, an Indigenous settlement of Seediq located in central Taiwan, as a case. Weaving in the Zhongyuan settlement is currently a distinct historical tradition, and this study focuses on the weaving process as a link between the settlement's business operations and the functioning of the SEPL. Therefore, the actions of the Weaver's Home (WH), the only entity that considers weaving as a cultural industry (which can also be considered the most active representative of the ethnic group) is the focus of this study.

3.1 Development of Weaving Techniques in Zhongyuan Settlement

Seta Iban (1919–2007), who was recognized as the most skilled weaver among the Seediq people, lived in Zhongyuan settlement all her life and taught many people to weave cloth. Thus, the settlement still has many weavers, including Seta Bakan (born in 1957), the granddaughter of Seta Iban, who has been registered as a national treasure by the State. Lin Ximei and Gao Xuezhu, also students taught by Seta Iban, have been registered as preservers of the craft by the local government and have conducted craft inheritance courses in the settlement. At present, the Zhongyuan settlement is proud to have the largest number of weaving culture preservers. Three of the four Seediq people registered by the country as preservers of weaving technology under the Cultural Assets Act live in the tribe. Weaving requires great endurance and the spirit of self-improvement, and social norms drive women to be good at weaving. From interviews in the settlement, the research team could vaguely feel that the settlement people are proud of themselves for coming from the Balan Community, and also have standards and expectations for their own abilities, as well as a spirit of not being defeated. These sentiments explain why the settlement has more weaving talents.

At present, the development of weaving in the Zhongyuan settlement can be divided into three types: the cultural asset preservation and skill transmission type, the cultural industry type, and the community development type, which features weaving as community development. The first type encompasses the preservers of the three skills mentioned above. They have government support and funds for the inheritance of skills, and their method is teaching ethnic minorities through systematic courses, focusing on acquiring skills and not involving commercial development. This practice has been promoted for 2 years with a total of about 20 students. The other two types are the cultural industry type, featuring product sales and cultural experiences, and the community development type, featuring weaving craft settlements. These two types are more relevant to the operation of the cultural business mechanism discussed in this article. The main promoter, WH, is the only entity in the settlement that operates weaving as a cultural industry, so this study discusses

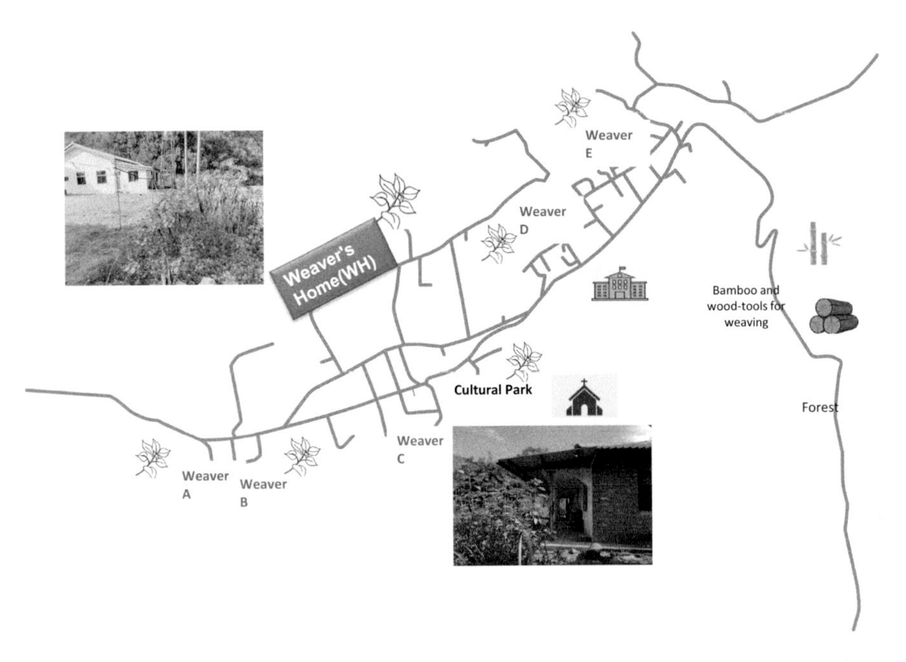

Fig. 9.7 Zhongyuan settlement: Ramie fields reappear in tribes for cultural use. (Source: Photo and drawing by Huang, Hsiu-Mei)

actions by WH. Regardless of the above types, weaver women have resumed the habit of planting ramie next to their homes, and ramie fields have also reappeared in the settlement (Fig. 9.7).

3.2 Weaver's Home (WH) Initiatives to Preserve Weaving Culture, Care for Livelihoods, and Promote Development

The founder of WH, Shuhui, was asked by her settlement people to set up WH in 2016. "*At that time, an aunt who knew how to weave said she wanted to pass on her skills, and said I was better at writing plans, so we tried to apply for a grant…*" (Shuhui 11/Jan/2020). The site, with an open space of about 400 m^2, provided a place for the staff to weave, and for visitors to shop and experience weaving, WH initially applied for the Ministry of Labor's employment counselling programme to support the livelihoods of settlement people by weaving, with government funds used to provide salaries. However, the programme can be operated commercially, and the ultimate goal is for the people to make a living on their own. Thus, there is a requirement for product output. "*It is very slow to weave cloth by hand, and it is difficult to meet the requirements.*" (ibid.) Thus, the workshop works on research and development of weaving products to attract customers and has diversified into settlement tours and weaving-related experiential courses to increase income.

Fig. 9.8 Weaver's Home (WH). (Photo taken by authors on 11 December 2021)

"*Because the programme has production requirements, the pressure is great… I also want to give up, but seeing their eyes eager for income, I don't know what to do!!*" (ibid.) After 6 years of implementing the government employment counselling programme, WH stopped applying. However, five of the staff remain, so more efforts must be made to seek financial resources for sustainable operation (Figs. 9.8, 9.9, and 9.10).

3.2.1 Analysis of WH Actions and Role in the Settlement

1. **Employment for livelihoods**

   As mentioned above, when WH was established in 2016, it started to provide livelihoods for the people of the settlement. Assistance from the government programme for staff salaries will not be received from 2023, and therefore WH must be more active in seeking financial resources to keep the workshop alive. "*Think about it, I am a daughter of the settlement, this workshop has been in the settlement for so many years, these settlement people are like family… Now everyone's income may have become less, but there is still some income to subsidize the family, and they are still willing to stay…*" (Shuhui 2022/12/26)

2. **Preservation of weaving and settlement culture**

   Inheritance of weaving skills is also one of the original motivations for WH's establishment. The government's employment counselling programme requires WH to guide tours based on the local history and weaving ecology. Therefore, over the years, even though government subsidies are decreasing and the pressure to survive is increasing, the original intention of preserving culture remains. The knowledge economy, such as teaching and demonstrating weaving, is supplementing income and contributing to the preservation of weaving culture.

3. **Culture-based weaving as a tiny business practice**

   UNESCO defines cultural and creative industries as: "*sectors of organised activity whose principal purpose is the production or reproduction, promotion, distribution and/or commercialisation of goods, services and activities of a cul-*

Fig. 9.9 Weaving products sold by WH. (Photo taken by authors on 11 December 2021)

Fig. 9.10 In the past, the tribespeople planted ramie next to their homes. The picture shows ramie replanted next to WH for craft experiences. (Photo taken by authors on 11 December 2021)

tural, artistic or heritage-related nature" (Europa Regina Creative Industries n.d.). According to Japan's Professor Miyazaki Kiyoshi (1996), in order for a community to be sustainable, residents must have secure jobs and incomes, making it necessary to develop local industries and provide jobs through the exploitation of unique local resources. Likewise, the improvement of business practices and the acquisition of sufficient funds are needed to promote the sustainable development of local self-sufficiency. The WH business model, which is a culture-based business practice, includes mainly the weaving of cloth by employees, the sewing of cultural goods such as bags and household items, and the introduction of a settlement tour kit that includes a DIY (do it yourself) weaving

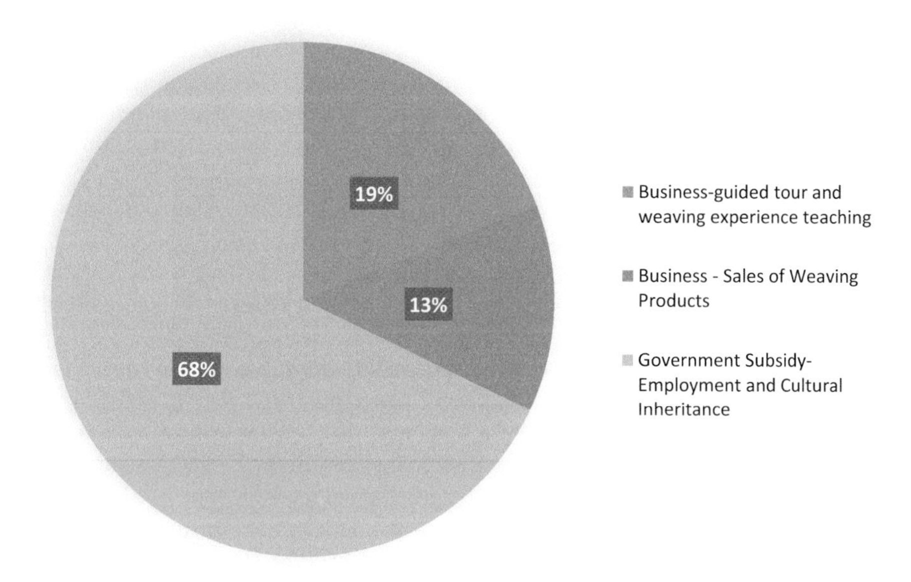

Fig. 9.11 Financial sources of Weaver's Home in 2022. (Source: Data provided by WH, redrawn by the authors)

experience. These contribute to livelihoods in the settlement and also serve to share Seediq culture. WH cooperates with external travel agencies and on-site camping areas that operate tribal package tours promoted by Facebook fan pages, which are extremely popular. However, despite these efforts, the commercial scale of the WH is still not large, and income still comes from government subsidies for cultural preservation and promotion. Taking 2022 as an example, government subsidies totalled about USD 3000, or 68% of the total income (Fig. 9.11). The remaining 32% is income from WH's product sales and cultural tours. Expenditures include paying the salaries of workshop employees and tribal people. The hourly fee is paid for guided tours and teaching of traditional skills. The person in charge of WH reported that these payments can barely be maintained at present.

3.2.2 A Small Force for Collaboration Between WH and YunTech: A Design Proposal for a Satoyama Spirit

Since 2020, the first author has led fieldwork for postgraduate students of a YunTech (National Yunlin University of Science and Technology) course. Each step in the course is a 5-week discussion and a 2-day/1-night fieldwork trip and design proposal for the Zhongyuan settlement. Such work has been done five times to date. In addition to WH itself, the cooperation partners have also been extended to tribal community development association members, hunters, women weavers, elders, and so on. Aims are to assist in identifying settlement issues and organizing the settlement culture knowledge system, and to try out solutions. Collaboration on the settlement design workshop is divided into three areas: (1) assisting in design pattern innovation for weaving at WH, (2) organising the knowledge system of weaving culture in the Zhongyuan settlement, and (3) exploring other settlement resources and organizing knowledge. The latter two are resource research and design proposals in the context of SEPL. For example, the students first study and understand the knowledge of ramie planting and experience the process of ramie production. They then organize their knowledge into knowledge content that is easily understood by the people. The results completed by students are fed back to the settlement for reference. "*In addition to trying to weave the weaving patterns designed by the students, we also display their reports in the workshop. Our workshop has no doors, so both tribespeople and outsiders can come in and out freely. We will see the results of our cooperation with universities.*" (Xiaoping 10/Jan/2023) (Figs. 9.12 and 9.13).

In addition, the team also conducted settlement resource surveys, paying particular attention to the interaction between environmental resources and residents' lives. For example, a survey was conducted on plants, hunting, and the diet of the ethnic group. The team also invited botanical experts to lead the students into the mountains and forests to gain local knowledge on how the settlement people identify and view plants. For example, hunters can judge the whereabouts of wild boars based on their footprints and signs of their eating the roots of the giant elephant ear plant and the fruits of banyan trees, and can then set traps accordingly. At the same time, surveys were used to observe and record the worldview and social norms of the tribespeople. For example, norms for hunters specify if a hunter catches a wild boar, he needs to go home. As long as he has enough to eat, he must not be greedy. When going out to hunt, if he sees the Formosan white-eyed nun babbler (The tribesmen call it the spiritual bird.) flying in the wrong direction, he should not go up the mountain; if you hunt a wild boar, others can brag about it, but you cannot do so yourself. Surveys also found that the tribe's eating utensils are made from local

Fig. 9.12 Students learning ramie scraping from tribal people in a workshop. (Photo taken by authors on 27 October 2022)

Fig. 9.13 Students help WH design new patterns, handed over to the tribe to weave and make bags for sale. (Photo taken by authors on 20 March 2023)

Fig. 9.14 Hunter demonstrates trap making for students. (Photo taken by authors on 23 April 2022)

materials and recorded simple cooking and sharing habits. In particular, when hunters hunt down a wild boar, they will definitely give it to the elders of the tribe.

From the investigation, the team found that the Zhongyuan settlement has a rich ecosystem. Traditionally, the tribespeople lived and interacted closely with the mountains and forests. Their norms and taboos allowed the environment to be maintained without over-hunting and balance to be maintained in the ecology. Students' investigations have included sorting systematic easy-to-read information to feed back to the tribe for reference that can be used as teaching material for tribal experience guides. One hunter told the team, "*The older generation is very familiar with the mountains and forests, but now that children no longer hunt, they are not familiar with mountain forest knowledge and the functions of plants. There is a lack of understanding and knowledge of animal habits. You helped us organize it into information. Now when I go out to give tours, I will keep it with me for easy reference.*" (Buhu 12/Sep/2023) (Figs. 9.14 and 9.15).

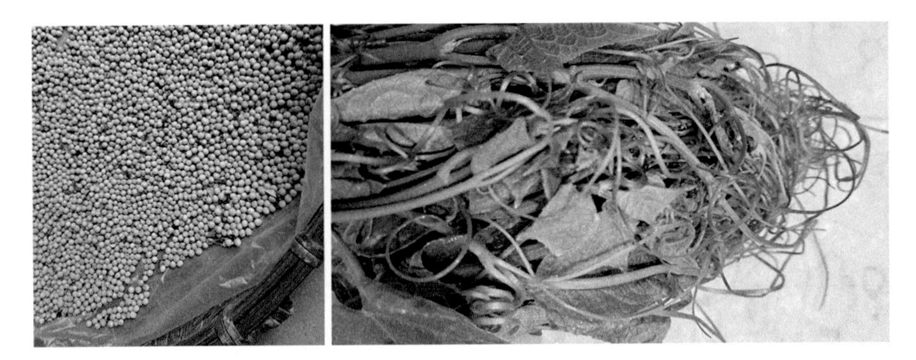

Fig. 9.15 Tree beans (left) and young chayote leaves (right) are common foods eaten by the tribe. (Photo taken by authors on 11 December 2021)

3.3 Limitations and Adjustments in the Business Mechanism of WH Weaving

3.3.1 High Labour Costs of Slow Manual Weaving

In the case of WH weaving, for example, one piece of cloth takes an average of about 3 days of work to weave. If innovative patterns are woven, it takes even more time. Thus, the resulting cultural products, such as bags, cannot be sold at low prices. In the capital market, it is difficult to compete with mass produced goods, or even cheaper manually produced goods from abroad. This is a general limitation for sales of traditional craft products.

3.3.2 Cultural Preservation and Ethnic Identity Are Difficult to Operate on a Profit-Making Basis

The intention of WH's operation is not necessarily purely to make money, "*My idea is not to be like another settlement agricultural product processor who also applies for employment counselling, where employees are paid on a piece-rate basis to meet the government's required output. I think we are a family and collaborators. We should work together for the survival of the workshop, adhere to handwork, and find out cultural characteristics, this is also the direction of efforts*" (Shuhui 26/Dec/2022). As Miyazaki (1996, pp. 27–32) said, "The community industry is based on the living culture passed down from generation to generation, the traditional wisdom of harmonious coexistence with nature, the beauty and ingenuity cultivated by the community in its own life, and the search for community." WH develops weaving as a tribal craft. Its goal is not just economic output or industrial revitalization of the community, but also the sustainable development of community culture through weaving. Therefore, it does not operate on pure business logic that reduces costs and puts profits first. As such, it is difficult to fully support the survival of the workshop by purely making profits from selling goods.

3.3.3 Responses and Adjustments

1. **Diversification and development of financial resources for WH internal strengthening and flexible external services**
   Examples include actively developing new products internally, setting up the workshop environment, working with settlement hunters and nearby food and accommodation businesses to promote guided tours on settlement culture and weaving experiences, and planning 2-day/1-night kit tours. Provision of external services should also be examined, such as participating in various markets and selling products online, collaborating with nearby camping areas to teach on-site weaving experiences, collaborating with primary and secondary schools in the county to teach about weaving, and collaborating with our university team to design work camps.
2. **Public value of WH actions as cultural preservation**
   Harvey (2008) considered that the "small heritage" of local residents sometimes becomes a counterweight to the larger discourse and forms a part of a collective image of the future. In the case of WH, its expectations involve not just making money and profit. As is the case for other small-scale community crafts in Taiwan that are not yet profitable in terms of industrial output, what they produce is social value rather than economic value (Chiang 2016). This social value includes the livelihoods of the settlement people, cultural patterns of life passed from generation to generation, and traditional wisdom for harmonious coexistence with nature. It involves everything from the cultural heritage of weaving to business operations, to the investigation of settlement culture and biodiversity. WH conveys the environmental knowledge and intelligence of the settlement through guiding and experiences. These activities embody the spirit of Satoyama and have social and public value. As Taiwanese society has not yet formed a habit for consuming local culture today, funding support from government departments has its own legitimacy. Based on the public value of cultural preservation and contribution to the satoyama, WH is looking for support from various government departments for programmes. These include training workshop employees to improve their weaving skills, creating cultural and historical records and navigation skills, and the teaching of weaving in primary and secondary schools.
3. **The construction of a weaving craft village—attracting residents' participation through cultural preservation and solidarity economy**
   Starting from 2023, WH will use the weaving craft settlement as its vision and seek financial support from the Ministry of Culture of the central government. The content of activities will mainly include the establishment and promotion of weaving knowledge systems, youth ethnic group training, investigation of plants and hunting stories in settlements, and the settlement solidarity econ-

omy. Cooperative planning will be carried out towards local autonomy and a sustainable community. The contents of the projects are to be planned by the tribespeople. The tribespeople themselves will teach the skills, tell and record their own stories, and make up the majority of participants. That is, taking the development of the entire tribe as the vision scale, actions will take into account both the economy and the environment (Fig. 9.16).

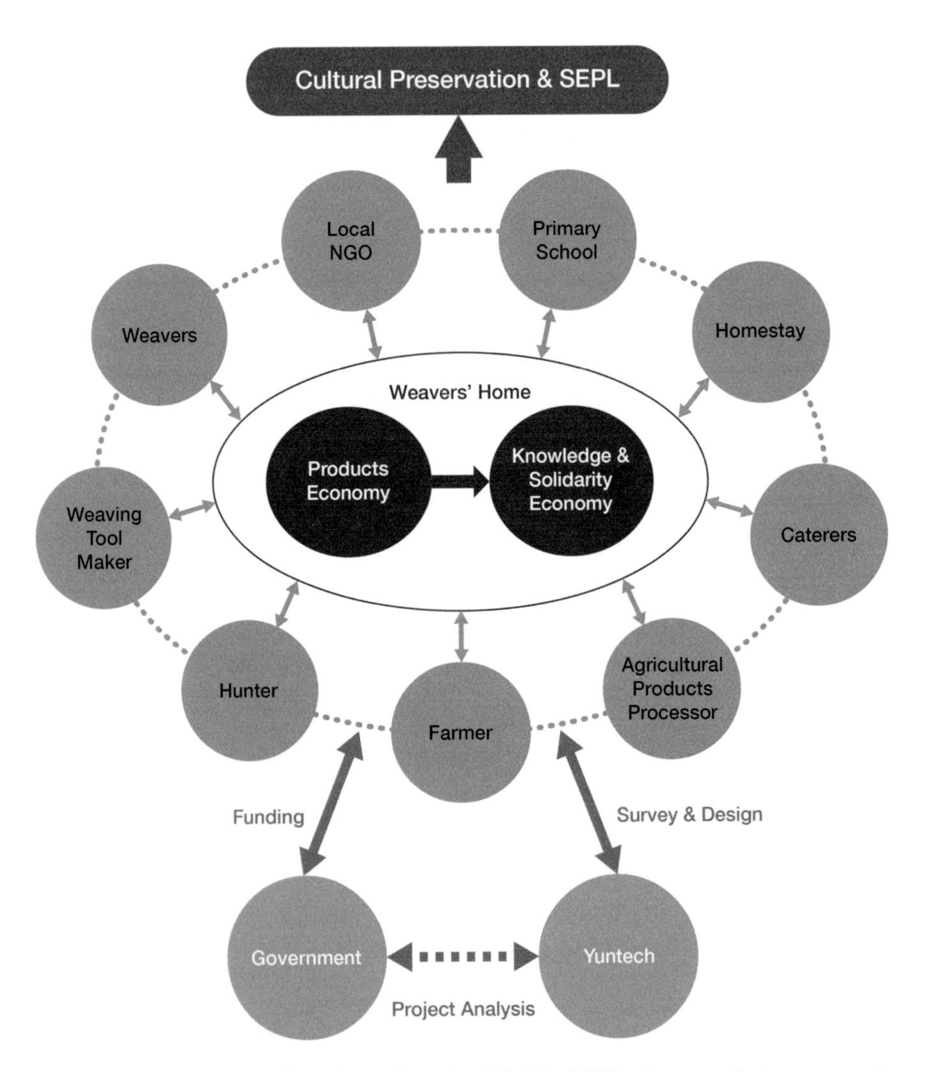

Fig. 9.16 Development and Business Operation Model of WH—From products economy to knowledge and solidarity economy. (Source: drawn by the authors)

3.4 WH Based on the Weaving Culture Business Mechanism Becomes a Protective Umbrella to Drive the Practice of Satoyama

Toledo and Barrera-Bassols (2008) point out that Indigenous peoples maintain and/or possess at least 80% of the planet's cultural diversity. They also posit that the presence, permanence, and resistance of these communities in ancestral territories have been fundamental for the conservation of ecosystems. These Indigenous territories can be considered socio-ecological production landscapes and seascapes (SEPLS) as they allow for a socio-ecological balance in the productive use of landscapes based on biocultural memory (Quintero-Angel et al. 2022). However, Taiwan's Indigenous people still face the difficulties of living in remote areas, poor transportation, few job opportunities, population emigration, and economic disadvantage. Therefore, settlement economic development is often a focus for both the government and the private sector, and settlements often seek tourism as a pathway for settlement industrial development because of their beautiful mountains and water. The Zhongyuan settlement has the same problems as most Indigenous settlements, so they have also looked to the settlement's characteristics for development projects. The research team believes that the settlement's restoration of the small-scale ramie landscape and weaving technique settlement is cultural capital as a "local development source." The settlement has natural capital in its rich ecosystem. Likewise, the planning and proper management of business mechanisms based on weaving culture revitalization and settlement tours can turn into economic capital to attract the participation of the settlement people. As a result, the small ramie field landscape and the inheritance of weaving skills, as well as the ecological knowledge of animals and plants, can be restored. Therefore, the two cultural business mechanisms of weaving culture revitalization and settlement tours can become the protective umbrella of the SEPL.

4 Conclusion

The well-maintained SEPL in a satoyama is the result of the harmonious coexistence of humans and nature. However, landscapes are not static, but are affected by people's daily actions and attitudes. Therefore, the action process and operational mode are key factors in maintaining the harmonious coexistence of humans and nature. The traditional weaving of the Seediq, from the process itself to the social interactions it encompasses, embodies this very close relationship between nature and society. That is, the landscape created to meet living needs (such as clothes and quilts) can be considered as a form of socio-ecological production landscapes and seascapes (SEPLS), allowing for a socio-ecological balance in the production and use of the landscape based on bio-cultural memory. Facing lack of need for weaving in life today, the cultural characteristics of the weaving ethnic group are being re-promoted as a cultural industry concept. The diverse development of WH-based weaving cultural capital of the Zhongyuan settlement, paired with the understanding and use of natural capital, has become economic capital for settlement livelihood development. In summary, the conclusions of this paper are described as follows:

1. From the perspective of the relationship between weaving skills and the environment, the weaving knowledge system involves the geographical environment, tools, individual skills, teamwork, and social network relations, and exhibits the wisdom of humans in response to their natural environment. This settlement has more weavers than other tribes, as well as those who can make weaving tools. The surrounding environment is rich in biodiversity, and the weaving operations continue to expand, giving the community an advantage.
2. From the perspective of the environmental landscape, the homestead ramie field serves different purposes in the past and present. The ramie field of the Indigenous ethnic group is a homestead ramie field for internal use. Ramie was planted to meet the needs of life in the past. Where there is a woman weaver, there is ramie field. Now, the settlement is reviving itself with weaving culture. Whether for the inheritance of cultural skills, the economic production of cultural industry, or the development of community characteristics, all actors will replant ramie near their homes for teaching or cultural experience, the same as the old generation.
3. Traditional weaving is limited by slow manual speed, high labour costs, and the mission of cultural preservation. Methods of adjustment include developing financial resources through diversified development, including knowledge economy and solidarity economy, cooperation with the settlement and nearby businesses, and winning financial support from the government based on the public value of cultural preservation and satoyama contribution. In addition to the development of weaving products, cultural tourism as a cultural management mechanism with rich ecological environment knowledge and weaving DIY experiences, will also be developed.
4. Cooperation between the university and the settlement plays a small role in the content of the proposals on the Satoyama spirit. The university enters the settlement to discover the settlement culture and biological diversity and assists in resource investigation and proposals related to the concept of community design. These include the collation of weaving knowledge, the investigation of plants valued by the ethnic group, and the investigation of hunters' knowledge of mountain forests and hunting, based on the concept of SEPL used in this project.
5. The revitalization of the tribe's weaving culture is based on the unique craftsmanship of the weaving ethnic group and the tribe's guided tours. These two cultural business mechanisms can become incentives for the tribespeople to participate. In order for these cultural business mechanisms to continue to operate, the tribespeople must start from understanding tribal history and weaving culture, as well as maintaining biodiversity. Good management is required to be sustainable. The research team concluded that these incentives form the umbrella that drives the SEPL (Fig. 9.17).

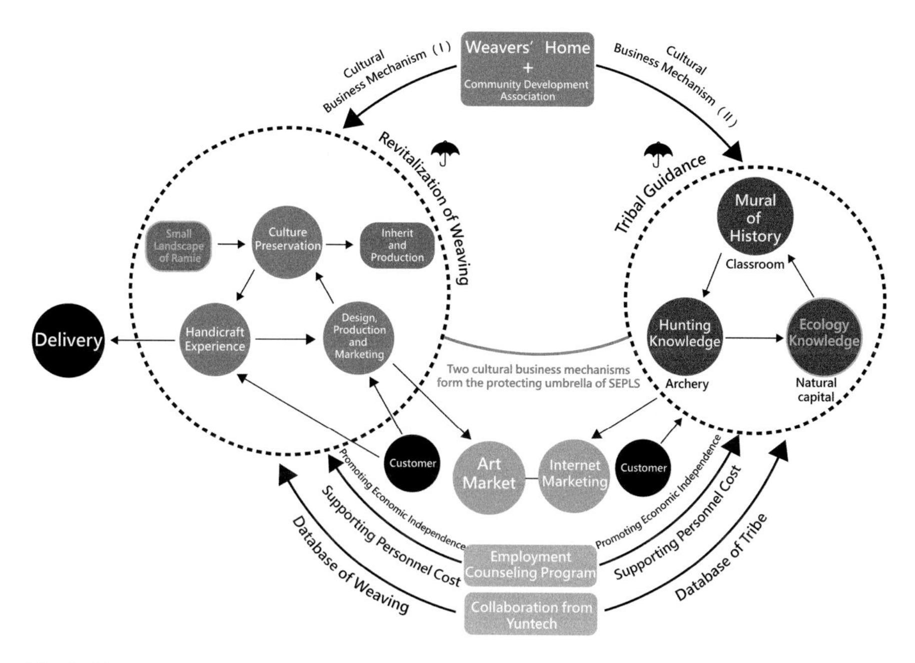

Fig. 9.17 Two kinds of cultural business mechanisms of Weaver's Home (WH). (Source: drawn by the authors)

Figure 9.17 depicts two kinds of cultural business mechanisms of the WH. On the left is route one: reviving the weaving culture and becoming the protective umbrella of the SEPL. Actions include weaving-related product design and marketing, weaving experience DIY, skill inheritance, and other actions, as well as the ramie field formed by these actions. On the right is route two: tribal guides can also become the protective umbrella of the SEPL. Actions include guided demonstrations of hunting knowledge, guided tours of tribal migration, legendary stories drawn by tribal people, and guided tours of ecological knowledge. The picture below is an explanation: the above two business mechanisms, whether the sale of goods or the income from tribal tour guides and skills inheritance experiences, allow WH to hire manpower, support the preservation of weaving culture, and expand the participation of tribespeople, promoting a tribal solidarity economy. This also changes the landscape. For example, ramie fields, weaving-related dye plants, and materials required for weaving tools were scattered among the tribes. Therefore, the business model mechanism based on cultural characteristics became an inducement for the tribespeople to participate and promoted the practice of a SEPL.

References

Cao LF (2021) Three Generations of Weavers: Skills and Memories of a Seediq Family. Nantou County Cultural Bureau (in Chinese)

Chen ZPL (2009) Some Reflections on the Study of Indigenous Peoples' Knowledge. Taiwan Indigenous Peoples Series 5:25–53 (in Chinese)

Chiang MC (2016) Crafting community development. Journal of Cultural Heritage Conservation 36:37–53 (in Chinese)

Europa Regina Creative Industries (n.d.) UNESCO Cultural and Creative Industries. Retrieved from https://europaregina.eu/organizations/igos/united-nations/unesco/

Google (2023) Google maps of Zhongyuan settlement, Ren'ai Township, Nantou County, Taiwan. https://www.google.com.tw/maps/place/546%E5%8D%97%E6%8A%95%E7%B8%A3%E4%BB%81%E6%84%9B%E9%84%89%E4%BA%92%E5%8A%A9%E6%9D%91/@24.0732282,120.9657012,3791m/data=!3m1!1e3!4m6!3m5!1s0x3468de74141408bd:0x9e563665a5199b0c!8m2!3d24.0746122!4d120.9589496!16s%2Fg%2F1q5bm2zn5?hl=zh-TW. Accessed 02 February 2023

Goss J (2005) The Souvenir and Sacrifice in the Tourist Mode of Consumption. In: Lew AA & Cartier C (eds) Seductions of Place: Geographical Perspectives on Globalization and Touristed Landscapes. London & New York, Routledge, pp 56–71

Harvey DC (2008) The History of Heritage. In: Graham B & Howard P (eds) The Ashgate Research Companion to Heritage and Identity. Hampshire & Burlington, Ashgate Publisher, pp 19–36

Ho WJ (2010) Weaving the rainbow-The case study of Seediq culture contexts and development about weaving handcraft at Alang-Gluban. Master's thesis, Yunlin University of Science and Technology, Yunlin (in Chinese)

Hsieh SC (2004) A Macro-exploration of Ethnic Anthropology: An Anthology of Taiwan's Aborigines. Taiwan University Press (in Chinese)

Hu TL (1993) Lanyu Perspectives. Institute of Ethnology, Academia Sinica, Taipei (in Chinese)

Hwang SH, Huang HM (2019) Cultural Ecosystem of the Seediq's Traditional Weaving Techniques—A Comparison of the Learning Differences Between Urban and Indigenous Communities. Sustainability 11(6):1519

Iwan PL (2014) Investigation on the Traditional Weaving Skills and Cultural Asset of Puniri of the Seediq, Cultural Asset Bureau, Ministry of Culture, Taiwan (in Chinese)

Liao HY, Tseng YH, Tzeng HY (2011) Ethnobotany on Seediq of Jungyuan Tribe in Nantou County, Taiwan. Quarterly Journal of Forest Research 33(1):17–34 (in Chinese)

Liu TR, Su HJ (1983) Ecology of Forest Plants. The Commercial Press, Ltd., Taipei, Taiwan, pp 94–97 (in Chinese)

Miyazaki K (1996) Community empowerment for a New Look. In: The Mind of the People: Ideas and Examples of Community Building in Japan. Handicraft Research Institute, Nantou County, Taiwan, pp 27–32 (in Chinese)

Nantou County Government (2007) Nantou Agricultural Land Resources Spatial Planning Plan Report. chrome-extension://efaidnbmnnnibpcajpcglclefindmkaj/https://www.tcd.gov.tw/uploadfile/region/1060724%E6%9C%9F%E6%9C%AB%E7%B8%BD%E7%B5%90%E5%A0%B1%E5%91%8A%E6%9B (in Chinese)

Quintero-Angel A, López-Rosada A, Quintero-Angel M, Quintero-Angel D, Mendoza-Salazar D, Rodríguez-Díaz SC, Orjuela-Salazar S (2022). Linking Biocultural Memory Conservation and Human Well-Being in Indigenous Socio-Ecological Production Landscapes in the Colombian Pacific Region. In: Nishi M, Subramanian S & Gupta H (eds) Biodiversity-Health-Sustainability Nexus in Socio-Ecological Production Landscapes and Seascapes (SEPLS). (Satoyama initiative thematic review), vol 7. Springer, pp 35–59

Shen MR (2008) The chorography of Ren'ai township. Office of Nantou County (in Chinese)

Tian ZY (2001) Taiwan's Indigenous People-The Atayal. Taiyuan Press (in Chinese)

Toledo VM, Barrera-Bassols N (2008) La memoria biocultural: la importancia ecológica de las sabidurías tradicionales. Icaria

Tseng LF (2013) A Pilot Study on Transmission of the Seediq Weaving Tradition and Handicraft Puniri in Nantou County: The Three Generations of Seta Iban as Research Subject. Journal of Cultural Heritage Conservation 24:33–57 (in Chinese)

Wan YY (2009) Craft of the Indigenous People-Aesthetic Expression and Possibilities of Contemporary Art. Taiwan Craft Quarterly 32:8–11 (in Chinese)

Wang MH (2003) Exploration on the Social Nature of the Atayal from the Perspective of Multiple Meanings of Gaya, Taiwan Journal of Anthropology 1(1):77–104 (in Chinese)

Wang MH, Iwan PL (2012) Activating the Seediq Culture: A Study of Cultural Industry in a Seediq Community. The Journal of Chinese Ritual, Theatre and Folklore 176:233 (in Chinese)

Wang SS (2001) Contemporary Art of the Indigenous People. National Taiwan Museum of Art Education, Taipei (in Chinese)

Yeh HY (2011) Tribe Skills/Memory and Cultural Industry: (Re)production of the Culture of Female Weavers of the Tailuge, Xiulin Township, 2011 Annual Meeting of Association for Cultural Studies (in Chinese)

Shyh-Huei Hwang Distinguished Professor and Library Director at National Yunlin University of Science and Technology (YunTech), Chinese Taipei.

Hsiu-Mei Huang Doctoral student at Graduate School of Design, National Yunlin University of Science and Technology, Chinese Taipei.

Tzu-Hsuan Chan Has a Master's degree from the Graduate Institute of Architecture and Cultural Heritage, Taipei National University of the Arts.

Chapter 10
Protecting the Rights and Livelihoods of Local Communities in the Face of Agro-Industrial Development to Conserve Socio-Ecological Production Landscapes (SEPLs): The Experience of Green Development Advocates (GDA) in Campo-Cameroon

Mbole Véronique, Aristide Chacgom Fokam, Alain Fabrice Mfoulou Bonny, Carrele Floriane Nguena Mawamba, Stéphane Nzakou Tchakounte, Laetitia Musi Adjoffoin, Corine Linda Ehowe Issova, and Green Development Advocates

Abstract The municipality of Campo, comprising a forest area incorporated into the 776,202-hectare Campo-Ma'an Technical Operational Unit (TOU) in Cameroon, is dedicated to the implementation of several development projects such as new agro-industrial projects. The project discussed in this study is the establishment of a new oil palm plantation and oil processing units, which aimed to convert 60,000 hectares of forest into an oil palm plantation. The rationale for this project may have originated from the palm oil production deficit in Cameroon estimated at 180,000 ton.

This agro-industrial project, however, has had devastating effects on the environment, biodiversity, and the livelihoods of local communities who depend on the forest for food, firewood, and other resources. This case study recounts the actions of Green Development Advocates (GDA), a Cameroonian civil society organization (CSO) that initiated advocacy to defend the rights of impacted local communities. The approach involved mobilizing different stakeholders to take actions and carrying out analysis to highlight the illegalities and impacts of the project. Meanwhile, support was provided to the communities for various income-generating activities such as livestock rearing to limit these impacts and strengthen their resilience in the face of the drastic decrease in their personal and communal use of forest space.

The results obtained were encouraging, given that the area of the land to be used for the agro-industry project was significantly reduced to 40,000 hectares. While this was the case, the prospects are not all that bright as the amount of forest land

M. Véronique · A. C. Fokam · A. F. M. Bonny · C. F. N. Mawamba · S. N. Tchakounte · L. M. Adjoffoin · C. L. E. Issova · Green Development Advocates (✉)
Green Development Advocates, Tsinga, Yaounde, Cameroon

M. Nishi et al. (eds.), *Business and Biodiversity*, Satoyama Initiative Thematic Review, https://doi.org/10.1007/978-981-97-7574-3_10

given out is still significantly high. This case study reviews these facts and discusses perspectives for the project's ongoing establishment in the area.

Keywords Park · Local communities · Rights · Biodiversity · Agro-industry · Agro-enterprise · Agro-ecology

1 Introduction

Cameroon is located at the heart of the Congo Basin in Central Africa. Cameroon is one of several countries that make up the Basin, which has the second largest tropical forest in the world after the Amazon. Cameroon is often referred to as "Africa in miniature" because of its rich natural resources and very fertile soils and forests. It encompasses five agro-ecological zones (NOCC 2021), including the mono-modal rainfall forest zone (4.57 million hectares (ha)), the bimodal rainfall forest zone (16.58 million ha), the highlands (3.12 million ha), the Guinean high savannahs (12.31 million ha), and the Sudano-Sahelian zone (10.04 million ha). Cameroon is also rich in endemic species. Despite, this rich diversity of natural resources, the country is now more than ever faced with the challenge of sustainable use and participatory exploitation. The poor state of governance and human rights in the country exacerbates this challenge (Assembe-Mvondo 2009). This is likely why the development vision adopted in 2009 by the government opted to make Cameroon "an emerging country, democratic and united in its diversity" by 2035 through three stages (MINEPAT 2009). The first stage of implementation of the vision, from 2010 to 2020, was marked by the development of the Growth and Employment Strategy Paper (GESP), the revision of which led to the National Development Strategy 2020–2030 (NDS-30), which focuses on structural transformation and inclusive development. Implementation of the said vision has led to numerous large-scale infrastructure projects (e.g. a port, bridges, dams, roads, and motorways) but has focused above all on the exploitation of non-oil natural resources (Biya 2022) such as wood and mines and also on the establishment of so-called "second-generation agriculture" (Biya 2011) with the extension of existing agro-industry projects (palm oil and rubber) and the creation of new agro-industries such as the one established in Campo.

The municipality of Campo, host of the new Campo agro-industrial project, is part of the 776,202-hectare (ha) Campo-Ma'an Technical Operational Unit[1] (TOU) (Ed. Colfer 2005), which is dedicated to the implementation of several development projects. These include the establishment of new palm plantations and oil processing units in an agro-industrial project that aimed to convert 60,000 ha of forest into a plantation. The palm oil production deficit, estimated at 180,000 tons in Cameroon

[1]A unit established to ensure the sustainable management of forests and the preservation of ecosystems.

(Nkongho et al. 2014) seems to be the reason behind this project. Previous studies have stressed the massive involvement of elites in the development of oil palm plantations especially after the collapse of the National Rural Development Fund (FONADER)[2]-sponsored smallholder scheme that was designed for poor farmers and developed at the advent of the economic crisis in the early 1990s (Elong 2003; Obam and Elong 2011; Obam and Tchonang Goudjou 2011; Levang and Nkongho 2012). Given the high benefits that elites stood to gain, they wielded influence on information made available on new plantations to the public, particularly local communities. They have also worked closely with local and administrative authorities to repress individuals or civil society organizations such as Green Development Advocates (GDA) who are working hard to expose the impacts of such projects. The unavailability of information on the agro-industrial project, influence of elites, and poor accessibility (bad roads and poor telecommunication networks) to the project zone have been some of the challenges faced in carrying out advocacy.

This agro-industrial project would probably have gone through unnoticed if the area concerned was not of high ecological and environmental interest. Indeed, the district of Campo is home to two parks, the Manyange Na Elombo Campo National Marine Park and the Campo Ma'an National Park. The marine park is of crucial interest for the reproduction and perpetuation of rare species such as the large sea turtles that come onshore in the village of Ebodjé. The forest slotted for conversion in the framework of the agro-industrial project was not a degraded forest as the agro-enterprise claimed, but a virgin forest (Ottaviani and Chiaf 2021). In addition to this primordial consideration for the environment, the said forest is home to six communities of Indigenous Bagyeli peoples whose culture and survival are being seriously undermined by the establishment of the agro-industrial project given that the forest serves as a site for traditional rituals and a source of food and other raw materials (e.g. firewood, herbs) for them. This land grabbing greatly affects the rights and interests of the communities living near the project site (Abesha et al. 2022) and considerably modifies their way of life through the destruction of their main source of subsistence. It is in view of these negative impacts that GDA undertook the actions to mobilize and support the riparian communities in the defence of their rights, and build their resilience whilst improving their standard of living following the arrival of agro-enterprise in Campo. To this end, six of the several affected communities were selected for GDA's support: Nko'elon, Akak, Malaba, Nazareth, Mabiogo, and Ebodjé.

The main objective of this chapter is to share the experience of GDA in defending the rights, securing the spaces, and conserving the practices of Indigenous peoples and local communities living around areas allocated to the agro-industrial

---

[2] FONADER was a public capital company with legal personality and financial autonomy. Initially, the FONADER was an organ of the Ministry of Agriculture in Cameroon. It was charged with the following duties:

- Administration, storage and distribution of subsidized agricultural inputs
- Promotion and distribution of agricultural credit
- Financing and monitoring of certain development projects.

project in Campo. It also aims to draw lessons for the protection of socio-ecological production landscapes (SEPLs) and the fight against rural poverty in a context marked by a growing demand for agricultural land (Abesha et al. 2022) or for the exploitation of mining resources.

1.1 Description of the Campo Municipality

The municipality of Campo (Fig. 10.1) is located approximately 77 km from the town of Kribi (chief town of the South Region) and extends over 313,500 ha. It is made up of 17 villages, six of which are dominantly populated by the Indigenous Bagyeli people (Table 10.1), with the Iyassa, Mvaï, and Mabi ethnic groups dominating in the others. It is bordered in the north by the Kribi 1er, Akom 2, and Niété municipalities, to the east by the municipality of Ma'an (district of the Ntem Valley), to the west by the Atlantic Ocean, and to the south by the Ntem River and the Republic of Equatorial Guinea. The major activities carried out by the people are fishing, hunting, farming, and petty trading.

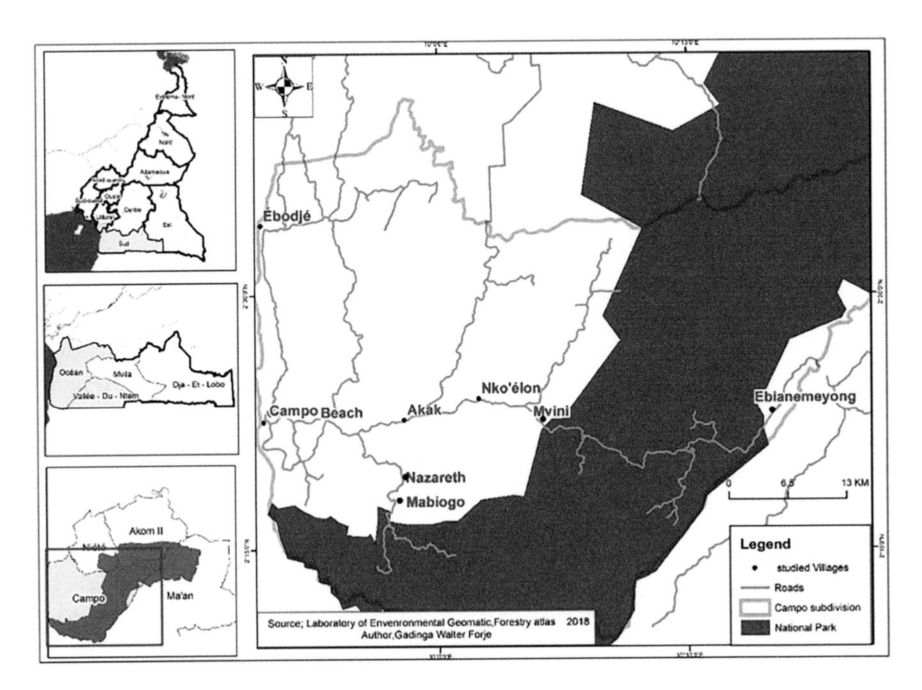

Fig. 10.1 Map of the Campo municipality. (Source: Forje G.W (2018) Forestry Atlas, Laboratory of Environmental Geomatics)

Table 10.1 Basic information of the study area

Country	Cameroon
Province	South
Division	Ocean
Municipality	Campo
Size of geographical area (hectare)	313,500 ha
Dominant ethnicity(ies), if appropriate	Bagyeli, Iyassa, Mvaï, and Mabi
Size of case study/project area (hectare)	313,500 ha
Dominant ethnicity in the project area	Bagyeli, Iyassa, Mvaï, and Mabi
Number of direct beneficiaries (people)	500
Number of indirect beneficiaries (people)	5032
Geographic coordinates (latitude, longitude)	2°21'59.99" N 9°48'59.99" E

Source: PCD 2014 based on the Plan Communal de Développement

Table 10.2 Timeline for the declassification and installation of agro-industrial project in Campo

Period	Activity
May 2019	Public Notice No. 0082/AP/MINFOF/DF/SDIAF/SC of the Ministry of Forests and Wildlife issued declassifying 88,147 ha of Forest Management Unit (FMU) 0925 located in the South Region
August 2019	Civil society statement on the agro-industrial case: A coalition of 40 CSOs led by GDA mobilized to denounce the irregularities of the declassification procedure and the negative impacts of the project on the area
September 2019	Public consultations in Campo and Niété conducted for the implementation of the project. The newspaper *Cameroon Business Today* reported on the agro-industrial case for the first time on 25 September 2019
November 2019	Prime Ministerial Decree 2019/4562 issued declassifying part of the FMU 0925. The decree concerned 60,000 ha of forest, i.e. two blocks measuring 40,000 and 20,000 ha respectively, for the creation of an industrial oil palm plantation in Campo and Niété
March 2020	Ministry of Forestry and Wildlife published a tender notice for the auction of 2500 ha of forest out of the 60,000 ha declassified
September 2020	First planting campaign launched by the agro-industry
March 2022	Presidential Decree No. 2022/112 issued allocating a provisional concession of 39,923 ha to the agro-industry: The area of 60,000 ha initially requested was reduced to 40,000 ha due to the CSO advocacy

2 Description of Activities (Methodology)

Within the context of assisting the communities in defending their rights, activities have been carried out by GDA in several phases (Table 10.2) over a period of 4 years (in 2020 and part of 2021 no actions were taken due to COVID-19) and are still ongoing, adopting a transdisciplinary approach. The methods and tools used for the study are presented below.

2.1 Data Collection

Classic methods of qualitative and quantitative data collection in the social sciences (Assembe-Mvondo et al. 2013) were used for this study. This involved a review of legislations on forest and land management with a particular focus on legislations in Cameroon. From September 2019, several field visits were conducted in both rainy and dry seasons in order to identify and work with six villages (Nko'elon, Akak, Malaba, Nazareth, Mabiogo, and Ebodjé) impacted by the establishment of the agro-industrial project. The villages were selected based on their willingness to work with our organization and their level of exposure to the potential impacts of the agro-industrial project. Focus group discussions and community meetings were organized from May 2021 to February 2022 in each of the communities in the presence of traditional and community leaders. The discussions covered the impacts of the establishment of the agro-industrial project on the environment as well as the livelihoods and cultural practices of the said communities. However, above all they focused on strategies to be implemented to halt the agro-industrial project. The tools used in the field, as described below give a clearer direction of the work done by GDA.

The defining characteristics of GDA's work lie in the mobilization of diverse parties and the tools used by the team for the protection of the Campo SEPL. These tools include the legal and environmental analysis of decrees on the release of land, participant observation, participatory mapping, and the development of historical lines and the seasonal calendar. The usefulness of these tools in advocating for the protection of the SEPL within GDA's working strategy is evident in the areas they impacted. Details are explained in subsequent sections.

2.2 Advocacy for the Non-Conversion of 60,000 Hectares of Land to an Oil Palm Plantation

In this context, the advocacy carried out was mainly based on in-depth analysis of the decrees on the release of the forest and mobilization of various stakeholders. The illegalities involved in setting up a modern oil palm plantation and processing unit in Campo were highlighted from the point of view of land tenure and forestry. With regard to the allocation of a 60,000-ha concession to the agro-enterprise, legal procedures were clearly not respected. The project site located in the municipalities of Campo and Niété was basically in the private domain of the state and had to be decommissioned before being granted to the agro-enterprise. The credibility of the entire process was a cause for concern given: the non-respect of the communities' rights to well-being; the communities' right to consent; and the fact that the environmental impact study was carried out after the decree on the release of the forest had been issued (GDA 2020). The provision of a plot of land to the agro-enterprise by the Minister of State Property and Land Tenure on 9 April 2020 before the provisional allocation decree by the President of the Republic, which was dated 22 March 2022, is a clear example of the irregularities of the Campo agro-industrial project. A timeline summarizing these irregularities is shown below (Table 10.2) and only adds to the discredit of the decried project.

The Campo Ma'an National Park is all the more important as it was created as an environmental compensation for the negative impacts of the Chad-Cameroon crude oil pipeline. The threats to the rich floristic and faunal diversity of the area (Wijma 2012) are undeniable, particularly for endemic species due to the destruction of their natural habitat. With regards to biodiversity, the impacts of the project are underscored in terms of the significant loss of species in the Campo area (GDA 2021a). In 2021, there were reports of two elephants being killed by unidentified persons around the village of Afang-Issoke, just months after the agro-enterprise began cutting down several ha (about 1500) of the forest. Rare species such as the Ebodjé sea turtles are also largely threatened. The traditional chief of the Ebodjé village during one of our visits stated that he "fears that the risks of pollution from the agro-industry will affect the nesting of this species which is the historic touristic specificity of village." The loss of non-timber forest products (NTFPs) necessary for the survival of the population is also a tragic consequence of the installation of the agro-industry for the local population (Nzakou 2020). In addition to the loss of protein sources, the most significant impact for them is the exponential increase in human-wildlife conflict (GDA 2021a). Culturally, the people also risk losing their ancestral worship sites in the forest, part of the reason they were able to maintain the biodiversity of the area over time.

2.2.1 Mobilizing of Various Actors Against the Installation of the Agro-Industrial Project

The actors mobilized in this context were of three types: the communities, civil society, and the media. The different actors each had a specific role to play in the advocacy campaign, at different times.

Community Mobilization

Community mobilization began during the first field visits to the area in 2019 and has subsequently continued as time goes on. In 2021, the communities were initially mobilized to map out and identify their living spaces and zones of high importance in a participatory manner. Using GPS, these communities recorded points for these zones that were then used by experts to develop maps. The maps (Fig. 10.2) revealed areas of high importance (for hunting, fishing, NTFPs, and sacred sites) that were

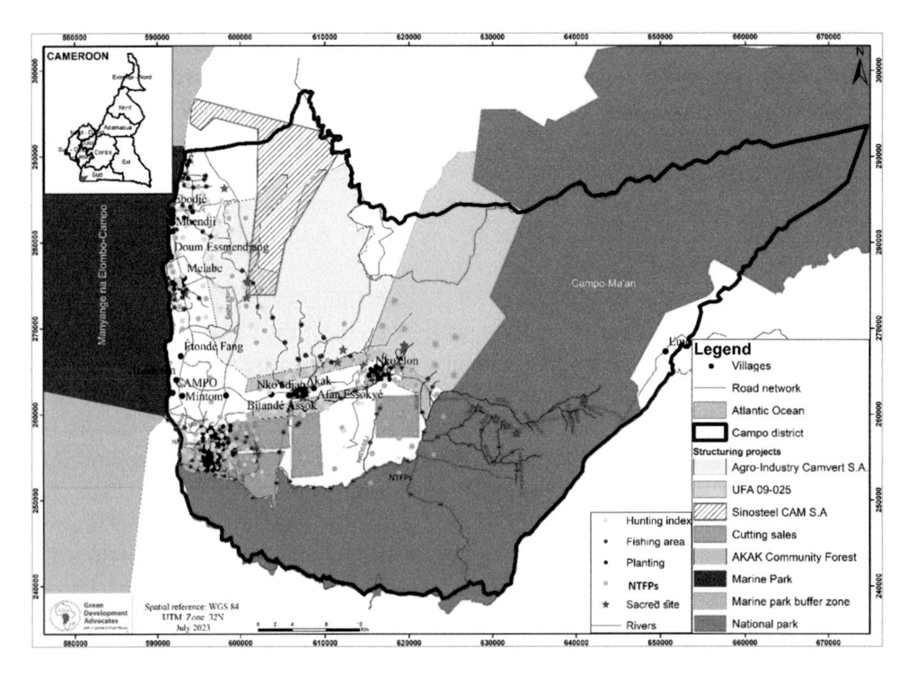

Fig. 10.2 Land-use and land cover map of case study site, geographical location of Campo Ma'an forest area. (Source: Prepared by author)

included in the concession allocated to the agro-enterprise and the Campo-Ma'an National Park confirming that the communities' fears were justified. This participatory map clearly demonstrated that these important areas were not taken into account for protection in the establishment of the modern oil palm plantation. Focus group discussions and historic timelines made it possible to identify key persons to take part in the advocacy. These elements made it possible to mobilize civil society to advocate against the risks associated with the agro-industry project.

Civil Society Mobilization

National and international civil society organizations (CSOs) were mobilized by GDA for the advocacy. At the national level, a coalition of 40 CSOs were mobilized through the signing of a communiqué in 2019 denouncing the negative impacts of the project on the ecologically sensitive area of Campo. More strategically, partnerships were formed with international organizations such as Global Witness, Synchronicity Earth, and Greenpeace Africa. The latter partnership led to the release of a damning report on the impacts of the agro-industrial project by Greenpeace and GDA in 2021 (GPA and GDA 2021) and a media campaign against the Campo agro-industrial project. Details of this campaign are summarized in publications on GDA's website: www.gdacameroon.org.

Media Mobilization

In 2021, traditional and social media were used to campaign against the establishment of the agro-industrial project in Campo. Precisely in September 2021, a field mission was organized to allow journalists to see the reality of the project's impacts (such as land conflicts, pollution, destruction, and loss of biodiversity) and to document cases of human-wildlife conflict created by the plantation's installation. Public opinion and press reports were able to reveal the gravity of the situation on the ground. Despite this damning information, the State allowed the project to go ahead. This led to the further mobilization of the media later in December 2021 for a press conference conducted in partnership with Greenpeace Africa.

The mobilization of all these actors led to meetings with the authorities. Notably at a meeting with the Minister of Forestry and Wildlife in February 2022, Greenpeace and GDA, along with a community leader, presented the results of an analysis and documentation on the project's impacts in the presence of the representatives of the agro-industrial project. The delay by authorities in responding to the advocacy carried out also permitted GDA to assist the local communities in dealing with some of the livelihood challenges they faced due to the project, as detailed in paragraphs below.

2.3 *Supporting Community Resilience to the Impacts of the Agro-Industrial Project*

Since the first planting campaign of the agro-industry project on 2500 ha of forest land, the local communities have been confronted with several impacts such as human-wildlife conflict, which have had grievous effects on their income-generating activities and cultural practices. Faced with these circumstances, GDA provided the communities with support for building resilience in their way of life. The activities carried out within this framework have made it possible for the communities to better organize, particularly the women, and to promote agro-ecological practices within the municipality.

2.3.1 Supporting Women in Building Resilience

During several meetings organized in the six target communities, emphasis was placed on sensitizing women to the need to structure themselves into a group. The aim was for them to wield more influence and to ensure their representation in decision-making processes, given that women suffer most from impacts related to such projects such as loss of farmlands. Based on this recognition, the women mobilized to form an association which had slowly been shaping up at the level of the Campo municipality, with sub-sections in each community. To strengthen the women's capacity, GDA facilitated their participation in knowledge exchange and experience sharing trips in August and October 2022. Themes were centred on advocating for the respect of women's rights, denouncing climate change impacts, and the promotion of agro-ecology. The extensive organization of women from other regions and countries motivated the women to extend the call for membership in their association to other villages within the municipality aside from the targeted six. Within this framework, in January 2023 GDA assisted the women in organizing and facilitating a workshop aimed at defining the fundamentals (e.g. governing body, objectives) upon which the women's organization would exist and function. Establishing this foundation has obviously made it easier to put in place further measures to assist them in carrying out agro-ecological practices and other income-generating and capacity-building activities.

2.3.2 Supporting Community-Led Agro-Ecological Practices and Building Resilience to Combat the Impacts of Human-Wildlife Conflict

Other activities within this framework were aimed at building the resilience of communities to the impacts caused by human-wildlife conflict. To achieve this, in 2021 a diagnosis of the communities' activities was conducted through participatory establishment of a seasonal calendar and an activity map. This allowed community members to rank and discuss any major activities that had been obstructed by human-wildlife conflict. Semi-structured questionnaires were then used to identify measures to be put in place to cope with human-wildlife conflict. Difficulties faced and the adaptation needs of the communities were defined in terms of agriculture and livestock rearing practices. In view of GDA's vision for green development, in 2022 an agreement was made with the communities, to engage in agro-ecology for the valourization of old fallow land. In its embryonic stage (SAILD and IRAD 2022) implementation of agro-ecology was not easy to set up. Thus, from March 2022, capacity-building trainings were organized on good agricultural and livestock rearing practices. Similarly, experience sharing and exchange trips on agro-ecology in August 2022 enabled the communities to better understand these concepts. The trainings were followed by the distribution of traditional seeds and native chickens in April and December 2022, respectively, aimed at the effective implementation of these agro-ecological activities such as intercropping and crop varietal mixtures.

3 Results

The results obtained from the activities carried out to protect the SEPLs around Campo differ from the goals established at the outset. That is, the results were mixed, with some positive albeit partially, and some not quite achieving the desired outcomes.

3.1 Reduction of Land Area Requested by Agro-Enterprise by 20,000 Hectares

The advocacy aimed at halting the establishment of agro-industry yielded positive results, which are outlined in subsequent paragraphs. The strategies were somewhat effective in that they led to a considerable reduction in the area of land requested for implementation of the agro-industrial project. Specifically, the 60,000 ha requested by the agro-enterprise was reduced. The Presidential Decree No. 2022/112 of 22 March 2022 allocated to the agro-enterprise an area of 39,923 ha for the implementation of its project, a difference of 20,077 ha. This decrease can principally be attributed to the advocacy strategies deployed by GDA.

3.2 Resilience to Human-Wildlife Conflict

Improvements to the resilience of communities to human-wildlife conflict have been minimal in that the implementation of agro-ecological activities has not effectively prevented human-wildlife conflict. Planted natural repellents such as pepper did not work to repel some of the animals. Similarly, the agro-ecological farmlands that were established continue to suffer from the devastating intrusion of animals expelled from their natural habitat due to clearing portions (2500 ha) of the forest for the agro-industrial plantation.

However, activities carried out have provided a framework for the reintroduction of some species of traditional crops such as maize, coco-yams, and yams that had disappeared from the communities. The introduction of plant species with high agro-ecological value, such as moringa, was accomplished. Similarly, the diversification of crops promoted by agro-ecological practices has allowed communities to be left with an acceptable portion of their crops. With regards to traditional chicken species (Fig. 10.3), the support provided has enabled the reintroduction of this species and boosted its rearing in the communities.

Fig. 10.3 Traditional chickens handed to the communities. (Source: Photo taken by authors, 2022)

4 Discussion

Regarding the resilience of communities to human-wildlife conflict, there has been an improvement over time. Their overdependence on dwindling NTFPs has been somewhat reduced. However, ongoing income-generating activities with the communities are aimed more at ameliorating their way of life in reaction to human-wildlife conflicts than anything else. Similarly, the ongoing implementation of agro-ecological projects has shown that even early in the activities, communities have a proper mastery of the good practices they learned in the trainings. It should be noted, however, that the support was provided in view of the advocacy carried out. Indeed, given the slow progress (3 years between the project's information note and the decree granting the provisional concession), there was a great risk that the communities would become weary and demobilized due to the lack of means of subsistence, or even give in to the manipulations of the agro-enterprise. However, the support provided helped to strengthen the credibility of GDA among the communities and to maintain their mobilization in the campaign against the installation of the project. As such, it is possible to suggest that a good advocacy strategy must be accompanied by concrete solutions to specific problems of communities, in addition to analysis and mobilization of various actors.

The participatory land-use maps produced as part of this advocacy work can be used going forward as a basis for demands that the communities' rights be better taken into account in the revision of the Campo-Ma'an National Park management plan and forestry policies. Similarly, the issue of land royalties owed to communities as a result of the concession granted to the agro-enterprise is a call for concern (Assembe-Mvondo et al. 2013). The failure to pay these royalties may be a subject of interest for the rest of the campaign, as well as the current exploitation of timber outside the area intended for the concession (GDA 2022).

With regard to the advocacy of GDA, it is worth noting that our consistent and non-violent approach towards the concerned stakeholders yielded visible results notwithstanding the long duration. Also, the strategy of mobilizing different actors permitted the campaign to gain more publicity and the attention of targeted persons. The mobilization of different stakeholders further permitted us to build the capacity of the communities and have them at the forefront of our campaigns. This involvement was visible in their engagement in media campaigns and workshops and the women's willingness to put in place an association. Our campaign with the communities also sparked other advocacy actions, leading GDA to produce a policy brief in August 2021 calling on the government of Cameroon to validate and implement a national oil palm strategy (GDA 2021b). The legal analysis also served as the basis for the joint submission of a complaint in 2023 to the United Nations Committee on the Elimination of Racial Discrimination (UN-CERD) by GDA, Forest Peoples Programme (FPP), and the Bagyeli Cultural and Development Association (BACUDA) and for other proceedings before the administrative courts brought by GDA.

5 Conclusion

Overall, the campaign to protect the SEPL in Campo has led to the reduction of 20,077 ha in the area of land requested by agro-enterprise. This result is attributed to the mobilization of various stakeholders and the legal and environmental analysis conducted to document the negative impacts of the establishment of this agro-enterprise. At the time of reporting, no other actions besides the media campaign had been taken by GDA for the protection of the Ebodjé sea turtles, though future actions are anticipated as activities in the area are ongoing.

It should be noted, however, that this project, although focused on advocacy for biodiversity and the rights and welfare of impacted communities, contributes to promoting the food and economic sovereignty of Cameroon. The campaign conducted was therefore aimed more at reducing the negative impacts of the agro-industry project on endangered species and local communities and ensuring that the rights of the local communities were taken into consideration. There remains a need to consider alternative ways of improving this project's effectiveness.

References

Abesha N, Assefa E, Petrova MA (2022) Large-scale agricultural investment in Ethiopia: Development, challenges and policy responses. Land Use Policy 117:106091

Assembe-Mvondo S (2009) Échec de l'État et gouvernance dans les États vulnérables: Une évaluation du droit forestier conformité et application au Cameroun. L'Afrique aujourd'hui 55(3):84–102

Assembe-Mvondo S, Brockhaus M, Lescuyere G (2013) Assessment of the Effectiveness, Efficiency and Equity of Benefit-Sharing Schemes under Large-Scale Agriculture: Lessons from Land Fees in Cameroon. Eur. J. Dev. Res. 25:641–656

Biya P (2011) Discours d'ouverture du comice agro-pastoral national d'Ebolowa, residency of the Republic, Republic of Cameroon, viewed 27 February 2023. Retrieved from https://www.prc.cm/fr/actualites/discours/1639-discours-du-president-de-la-republique-s-e-paul-biya-a-l-occasion-de-la-ceremonie-d-ouverture-du-comice-agro-pastoral-d-ebolowa

Biya P (2022) Message du chef de l'Etat à la nation à l'occasion de la fin d'année 2022 et du nouvel an 2023, Presidency of the Republic, Republic of Cameroon, viewed 27 February 2023. Retrieved from https://www.prc.cm/fr/multimedia/documents/9454-message-du-chef-de-l-etat-a-la-nation-a-l-occasion-de-la-fin-d-annee-2022-et-du-nouvel-an-2023#tabMediaRelated

Colfer CJP (ed) (2005) The Complex Forest: Communities, Uncertainty, and Adaptive Collaborative Management. Routledge

Elong JB (2003) Les plantations villageoises de palmier à huile de la Socapalm dans le bas Moungo (Cameroun): un projet mal integré aux préoccupations des paysans. Les Cahiers d'Outre Mer 224:401–418

Eno-Nku M (2021) Determinants of ecotourism development in and around protected areas: The case of Campo Ma'an National Park in Cameroon, Scientific African 11e00663, ISSN 2468-2276

Forje G.W. (2018) Forestry Atlas, Laboratory of Environmental Geomatic) Retrived from Forje GW, Tchamba MN,

Forje GW, Tchamba MN, Eno-Nku M (2021) Determinants of ecotourism development in and around protected areas: The case of Campo Ma'an National Park in Cameroon, Scientific African 11e00663, ISSN 2468-2276

GPA & GDA (2021) `Camvert: un cauchemar récurrent, Greempeace Africa & Green Development Advocates,viewed 8 December 2021. Retrived from http://gdacameroon.org/download/822/

GDA (2020)The "minor illegalities" of the process of declassification and concession of 60,000 ha of forest for thebenefit of an agro-industry in Campo and Niété, Green Development Advocates, viewed 4 September 2020. Retrived from http://gdacameroon.org/download/320/

GDA 2021a Degazetting and granting of 60,000 hectares of prime forest for oil palmproduction project in campo and nyete sub-divisions – ocean division, south region ofcameroon: Evaluating the biodiversity impact of the CAMVERT project on the Campo Ma'an TechnicalOperations Unit (TOU) , Green Development Advocates, viewed May 2021. Retrived fromhttps://gdacameroon.org/download/695/

GDA 2021b For a sustainable oil palm production in Cameroon: the imperative need to adopt and implement asustainable national strategy for palm oil production, Green Development Advocates, viewed August 2021. Retrivedfrom https://gdacameroon.org/download/744/

GDA, 2022, Analysis Note, Camouflaged destruction: Plundering of Campo Forests under the pretext of a socalled"Development" Project : Legal Analysis of Orders N0. 0011, 0012, 0013, 0014, 0015 Granting Timber Salesas Part of the Implementation of a Development Project in the National Domain in Campo, Green DevelopmentAdvocates, viewed 4 November 2022. Retrived from http://gdacameroon.org/download/893/

Levang P, Nkongho RN (2012) Elites et accaparement des terres au Cameroun: L'exemple du palmier à huile.Enjeux 47–48:67–74

MINEPAT (2009) Cameroun Vision 2035, Ministère de l'Economie, de la Planification et de l'Aménagement duTerritoire

Nkongho RN, Feintrenie L, Levang P (2014) The non-industrial palm oil sector in Cameroon. Working Paper 139.Center for International Forestry Research (CIFOR), Bogor, Indonesia
NOCC (2021) Agricultural calendar for the Five Agro-Ecological Zones of Cameroon. General Directorate,National Observatory on Climate Change Cameroon
Nzakou TS (2020) Dynamique du couvert forestier et impacts socio-environnementaux aux potentiels de laconversion des forêts du domaine permanent pour la production agro-industrielle dans les arrondissement deCampo et Niété, CRESA Msc
Obam FM, Elong JG (2011) Réponse Paysannes à l'expansion des Plantations de Palmier à huile dans la Région du Sud au Cameroun. In Elong J.G. (ed.) L'élite urbaine dans l'espace agricole africain. L'Harmattan, Paris 212–224
Obam FM, Tchonang Goudjou B (2011) Plantations de Palmiers à Huile des Elites Urbaines, Mutations Sociospatiales et Effets D'entrainement dans la Région du Sud (Cameroun). In: Elong JG (ed.) L'élite urbaine dans l'espace agricole africain. Paris: L'Harmattan. 200–209
Ottaviani J, Chiaf I (2021) Lungs of the Earth: Central Africa - Democratic Republic of Congo. Rainforest Journalism Fund, viewed 4 June 2021. Retrieved from https://rainforestjournalismfund.org/stories/lungs-earth-central-africa-democratic-republic-congo
PCD (2014) Plan Communal De Developpement De Campo, Commune de Campo.
SAILD & IRAD (2022) Cartographie des acteurs et des pratiques de l'agroécologie au Cameroun, Service d'Appui aux Initiatives Locales de Développement and Institut de Recherche Agricole pour le Développement, Rapport
Wijma OW (2012) Résumépublic de l'aménagement et de la démarche FSC pour l'UFA 09-025, Ministry of Forests and Wildlife (MINFOF), Yaoundé

Mbole Véronique Project Assistant at Green Development Advocates, Cameroon, working on projects focused on assessing the policies and impacts of large infrastructural projects on surrounding communities and assisting communities in agro-ecological and seed conservation practices.

Aristide Chacgom Fokam Coordinator and Environmental Jurist at Green Development Advocates, Cameroon, working on projects focused on assessing the impacts of policies on different actors, particularly communities around large infrastructural and agro-industrial projects and protected areas.

Alain Fabrice Mfoulou Bonny PhD candidate in public law and a Jurist at Green Development Advocates, Cameroon. Works on projects focused on assessing the impacts of policies on different actors, particularly communities around large infrastructural and agro-industrial projects.

Carrele Floriane Nguena Mawamba Project Officer at Green Development Advocates, Cameroon. Works on projects focused on mapping out community spaces and assessing the impacts of agro-industrial projects on surrounding forest communities and assisting communities in agro-ecological and seed conservation practices.

Stéphane Nzakou Tchakounte Project Assistant at Green Development Advocates, Cameroon. Works on projects focused on mapping out community spaces and assessing the impacts of agro-industrial projects on surrounding forest communities and assisting communities in agro-ecological and seed conservation practices.

Laetitia Musi Adjoffon Research Intern at Green Development Advocates, Cameroon. Works on projects focused on evaluating how ecosystem services affect and are affected by modified land use practices in the transformation of forest-agriculture boundaries.

Corine Linda Ehowe Issova Research Intern at Green Development Advocates, Cameroon. Works on projects focused on evaluating the well-being of forest communities around agro-industrial and large infrastructural projects.

BY NC SA

Chapter 11
Participatory Sustainable Production of Panela (Brown Sugarcane) and the Conservation of Tropical Dry and Very Dry (Subxerophytic) Forest in Colombia

Andrés Quintero-Ángel, Sebastian Orjuela-Salazar, Diana Saavedra-Zúñiga, Alejandro Castaño-Astudillo, María Viviana Borda Calvache, and Mauricio Quintero-Ángel

Abstract Dry and very dry tropical forests are some of the most threatened ecosystems in the world. In Colombia, they are in a critical state due to land cover and land use changes. Even though the soils are not very productive due to the harsh environmental conditions and scarcity of water, many of these ecosystems have served as the base of economic development in certain regions. Particularly, the subxerophytic enclave of Atuncela (SEA) in Colombia gradually became an agricultural zone with the expansion of sugarcane (*Saccharum officinarum*) cultivation. This shift has changed the landscape over the years, but high biodiversity in the remaining fragments still provides goods and services necessary for the inhabitants' well-being. Given the importance of these ecosystems as well as the high-pressure conditions exerted on them, in 2007, the regional environmental authority declared 1011.05 hectares of the SEA as a protected area. The aim of this case study was to present the direct relationship between the establishment of a protected area and goal-based planning that seeks to reduce threats by working to improve people's livelihoods. A survey was conducted for 120 properties to evaluate the ecosystem services offered. It was determined that 36.67% of the properties have areas of conserved forest, while 63.33% have crop or livestock areas. The economic activities have a small scale of commercialization and low technological intensity that meets acceptable

A. Quintero-Ángel (✉) · D. Saavedra-Zúñiga · A. Castaño-Astudillo · M. V. B. Calvache
Corporación Ambiental y Forestal del Pacifico – CORFOPAL,
Cali, Valle del Cauca, Colombia
e-mail: direccioncientifica@corfopal.org

S. Orjuela-Salazar
Fondo para la Acción Ambiental y la Niñez - Fondo Acción, Bogotá DC, Colombia

M. Quintero-Ángel
Universidad del Valle, seccional Palmira, Palmira, Valle del Cauca, Colombia

M. Nishi et al. (eds.), *Business and Biodiversity*, Satoyama Initiative Thematic Review, https://doi.org/10.1007/978-981-97-7574-3_11

living standards for the landowners. There is a sugarcane mill enterprise, where the community carries out technical and sustainable processing of panela and promotes the conservation of nature. The ecosystem services on which the community depends most are water regulation and soil formation. The inhabitants of this territory recognized the importance of natural resources for economic activities and as part of their territorial identity that favoured associative nature conservation.

Keywords Agriculture · Community sugar mills · Subxerophytic · Conservation · Protected area · Livelihoods

1 Introduction

Dry and very dry (subxerophytic) tropical forests are important for their high biodiversity and the fundamental ecosystem services they supply, such as water and climate regulation, carbon sequestration, soil retention, and legume species for human sustenance (Maass et al. 2005; Miles et al. 2006; WWF 2013; Blackie et al. 2014; DRYFLOR et al. 2016; Siyum 2020). However, these important ecosystems are some of the most threatened in the world (Miles et al. 2006; WWF 2013; Eds. Pizano and García 2014). In Colombia, the subxerophytic forests are in a critical state of fragmentation and deterioration due to changes in land use that have occurred since the twentieth century (Alvarado-Solano and Otero Ospina 2015; Arcila-Cardona et al. 2012; Etter et al. 2008; Etter and van Wyngaarden 2000; García et al. 2014). It is estimated that only 8% of the nine million hectares of dry and very dry forest that existed in the twentieth century in Colombia survive at the present time (García et al. 2014). They were replaced by pastures and agricultural fields and transformed by urbanization. The result was not only deforestation, but also desertification (García et al. 2014; Eds. Pizano and García 2014).

Even though the soils in subxerophytic ecosystems are not very productive due to harsh environmental conditions and scarcity of water, many of these ecosystems have been the indisputable axis of economic development in some regions where they are found (Maass et al. 2005; Kok and Alkemade 2014). In the Colombian case, one of the main economic activities developed in the dry ecosystems is the cultivation of sugarcane (*Saccharum officinarum*). Artisan production of panela (brown sugarcane) is an activity of great socio-economic importance for small producers, and many rural communities in Colombia derive their livelihoods from it (Rodríguez et al. 2020).

Panela production occurs in the subxerophytic enclave of Atuncela, a landscape which is part of a Key Biodiversity Area (COL 36), the dry enclave of the Dagua river basin (Critical Ecosystem Partnership Fund 2021), located on the western mountain range in Valle del Cauca, Colombia. This zone is a biodiversity-rich area that gradually became an area for agricultural enterprise with the expanded cultivation of sugarcane and the creation of artisan sugar mills (CORFOPAL and CVC 2022). As a result, this landscape has changed over the years, and the high biodiversity present in the remaining nature provides the inhabitants of the region with the goods and services they need for their well-being. Accordingly, we consider it a socio-ecological production landscape (SEPL).

According to the National Administrative Department of Statistics website (DANE 2023), sugarcane cultivation has tended towards monoculture, becoming the main factor of environmental change in Valle del Cauca, where it is the dominant feature of the landscape. This model of agricultural production has generated a very significant deterioration in different ecosystems and has significantly diminished the extent of very dry tropical forest (subxerophytic) ecosystems. However, producers in Atuncela have managed to stop the deterioration of the subxerophytic ecosystem based on the formation of an association of sugarcane mill enterprises and changes in their production and marketing practices. The inhabitants in the territory, thus the community, decided that the production of panela should no longer be done individually but in a community sugarcane mill managed by themselves, through the Paneleros Association of Atuncela—ASPAT (for its acronym in Spanish).

Because very dry tropical forest (subxerophytic) ecosystems are important for the conservation of endemic species and provision of ecosystem goods and services, in 2007 the regional environmental authority designated 1011.05 hectares of the subxerophytic enclave of Atuncela as an Integrated Management Regional District. This protected area category allows the inhabitants of the area to cultivate and carry out orderly and responsible livestock activities. In addition, the area became part of a network of protected areas in the region comprising the Farallones de Cali National Natural Park, three National Forest Protection Reserves (Anchicaya, San Cipriano, and Dagua), another Integrated Management Regional District (Chilcal), and the District of Soil Conservation of the Rio Grande. Currently, the majority of these conservation units are highly populated and transformed, hence functioning as isolated islands under different (and ineffective) management schemes (Orjuela-Salazar et al. 2018). The main objective of this case study was to present the direct relationship between the establishment of a protected area and goal-based planning that seeks to reduce threats by working to improve people's livelihoods. The study is also intended to demonstrate how the implementation of strategies focused on green businesses and sustainable production systems, such as the sugarcane mill enterprise, has had a positive impact on biodiversity and the well-being of inhabitants in general.

2 Methodology

2.1 Study Area: Integrated Management Regional District Subxerophytic Enclave of Atuncela (IMRDSEA)

The Integrated Management Regional District Subxerophytic Enclave of Atuncela (henceforth IMRDSEA) is located in the Corregimiento of Atuncela, Municipality of Dagua, Valle del Cauca, Colombia (Table 11.1, Fig. 11.1). It overlaps with the dry enclave of Dagua Key Biodiversity Area (KBA) (COL 36) (Fig. 11.1), within the Paraguas-Munchique-Bosques Montanos del Sur de Antioquia corridor of the Tropical Andes Hotspot (Critical Ecosystem Partnership Fund 2021). It is located at the altitudinal range of 600 to 1800 m above sea level, registers temperature values that oscillate between 22.66 and 33.76 °C, and precipitation values between 15.92 and 83.67 mm with a bimodal precipitation regime annually (CORFOPAL and CVC 2022).

IMRDSEA (Fig. 11.2) has 20 types of vegetation cover, where natural cover comprises 56.68% of the area, which consists of forests, streams, and shrubs in succession, among others. The rest, 43.32% of the area, corresponds to land covers that have been transformed or intended for production such as pastures for livestock, cropland, and houses (CORFOPAL and CVC 2022).

IMRDSEA has 155 recorded species of plants, 14 species of fish, 15 species of amphibians, 138 species of birds, 23 species of mammals, and 14 reptile species (Alexander von Humboldt Biological Resources Research Institute 2015; CORFOPAL and CVC 2022). Endangered species are endemics such as the Ruiz's robber frog (*Strabomantis ruizi*) and the Tlatepusco vanilla (*Vanilla odorata*). Species such as the Cauca poison frog (*Andinobates bombetes*), the northern tiger cat (*Leopardus tigrinus*), and the Colombian night monkey (*Aotus lemurinus*) are classified as vulnerable. Locally threatened species that are not globally threatened are the Cauca lily (*Eucharis caucana*), the Loboguerrero cactus (*Melocactus curvispinus* sub. *loboguerreroi*), and the Colombian red howler monkey (*Alouatta seniculus*).

Table 11.1 Basic information of the study area

Country	Colombia
Province	Valle del Cauca
Municipality	Dagua
Size of geographical area (hectare)	2334.37
Dominant ethnicity(ies), if appropriate	Atunceleños
Size of case study/project area (hectare)	2334.37
Dominant ethnicity in the project area	Atunceleños
Number of direct beneficiaries (people)	150
Number of indirect beneficiaries (people)	460
Geographic coordinates (latitude, longitude)	3° 46' 21.2484"N, −76° 39' 42.1272"E

Source: Authors' elaboration 2023

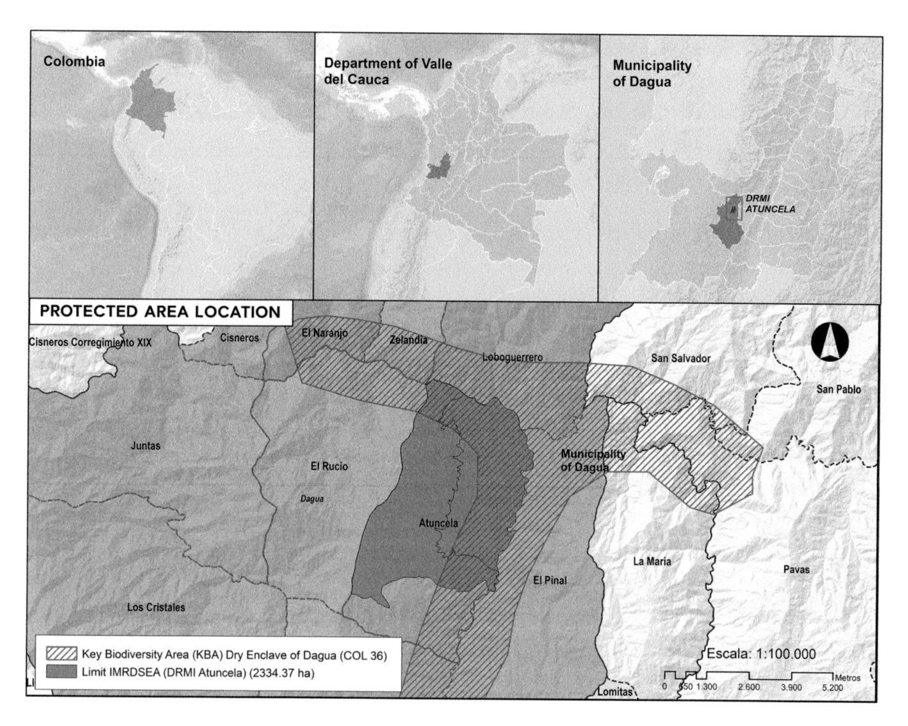

Fig. 11.1 Location of IMRDSEA in dark green (DRMI Atuncela by its acronym in Spanish). (Map: Authors' elaboration 2023)

Fig. 11.2 Panoramic view of the ecosystems in the landscape of the IMRDSEA. (Photo: Castaño 2022)

2.2 Information Gathering and Analysis

Information gathering and analysis was conducted during several meetings and workshops with the community between February and September 2022 in three phases (Fig. 11.3). Methodology is as follows:

1. Identification, characterization, and prioritization of stakeholders: This was carried out following the methodology proposed by Gómez et al. (2007). As a first step, a baseline for stakeholders and key data was established based on relevant documents and a review of literature on the 2007 declaration process and subsequent studies. Subsequently, workshops were held with the community to identify the stakeholders in the territory and characterize them taking into account changes in their numbers and identities from 2007 to 2022. Using the data obtained, stakeholders were classified into five categories: local, public, private, academic, and non-governmental. Furthermore, in these workshops, the stakeholders were prioritized in terms of their roles and responsibilities for managing the territory and the protected area within it, based on two exercises. In one exercise, attendees in two subgroups analysed actors by the following criteria: the role of each stakeholder, their scope of action, and what their relationship with the territory has been to date. Subgroup 1 analysed local, public, and private stakeholders, and subgroup 2 analysed non-governmental and academic stakeholders. The other exercise was conducted in the plenary. Using the analyses and classifications decided upon by each subgroup, stakeholders were classified into three types of stakeholders: (a) priority, (b) influential, and (c) allies. Finally, an analysis was carried out to categorize the relationship of the stakeholders with the IMRDSEA as local, regional, national, or international, and a prioritization was carried out according to the roles of the stakeholders.

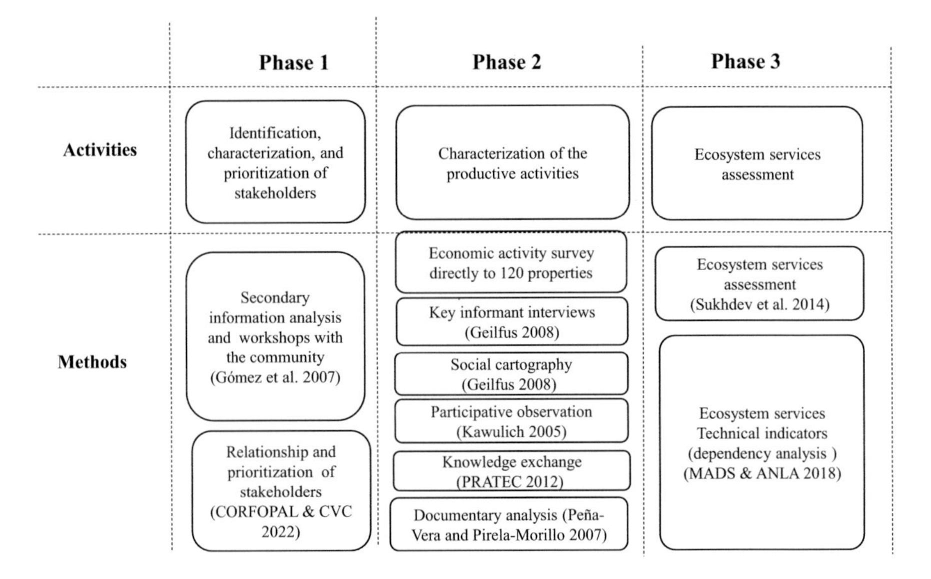

Fig. 11.3 Description of methods and activities

2. Characterization of productive activities: Information was collected using a variety of techniques including economic activity surveys on 120 properties, interviews with key informants (Geilfus 2008), participant observation (Kawulich 2005), social cartography (Geilfus 2008), documentary analysis (Peña-Vera and Pirela-Morillo 2007), and knowledge exchange (PRATEC 2012). Once information was collected, it was processed and systematized, and the corresponding qualitative and quantitative analysis was carried out. With the combined use of these methodologies, it was possible to characterize the productive activities of the inhabitants of the IMRDSEA and their relationships with the conservation of this SEPL.
3. Ecosystem services assessment: Following the methodology proposed by Sukhdev et al. (2014) and MADS & ANLA (2018), the state of ecosystem services was determined through the definition and use of technical indicators (quantitative and qualitative), as well as the analysis of the abiotic, biotic, and socio-economic factors. Once the ecosystem services were identified, a dependency analysis was conducted for the IMRDSEA in socio-economic and usage terms. This made it possible to establish the degree of importance of the service for the community in order to describe not only the identified services but also the interactions they present with the community, thus enabling prioritization of conservation actions. This dependency analysis consisted of the qualitative determination of the state of ecosystem services identified by analyzing the relationships between the natural environment and the community, classifying the services as (a) high, when fully functional and of good quality; (b) medium, when the service has been altered by human activities; and (c) low, when the service is practically non-existent. Once this classification was obtained, we quantitatively established, through co-working with the community, the level of dependency of the means of subsistence in three scores: (a) high dependency (3), when the means of subsistence of the community depend directly on the service; (b) medium dependency (2), when the community benefits from the service but its subsistence does not depend directly on it; and (c) low dependency (1), where the community benefits from the service but its subsistence does not directly or indirectly depend on it.

3 Results and Discussion

3.1 *Identification, Characterization, and Prioritization of Stakeholders in the IMRDSEA*

The following stakeholders were identified in the territory of the IMRDSEA: (a) local stakeholders who were the inhabitants of the territory, most of whom were descendants and heirs of the first settlers; (b) public stakeholders, including government authorities and representatives of the State; (c) private stakeholders, who were the companies that operated in the territory; (d) non-governmental stakeholders, who were recognized by the community for their actions and projects; and (e) academic stakeholders, mainly public and private universities (Table 11.2). The

Table 11.2 IMRDSEA stakeholders

Type of stakeholders	Description	Identified stakeholders	Scale
Local	They are the inhabitants and community organizations that reside and have their activities in Atuncela	Association of Producers of Atuncela—ASOPROCAT	Local
		Paneleros Association of Atuncela—ASPAT	Local
		Xerofítica entrepreneurship	Local
		Turisteando por Atuncela	Local
		Atuncela's Senior Group	Local
		Absentee owners	Local
		Community action board—JAC de Atuncela	Local
		Residents of Atuncela	Local
		Atuncela aqueduct users' association—ASUACAT	Local
		La Vigía aqueduct users' association—Aguavigía	Local
		Agricultural producers of Atuncela	Local
		Vista Hermosa village community	Local
Public	They are the entities that govern the territory, implement public policies, have power and influence on the issues of the protected area, or have a presence in the territory	Loboguerrero's Educational Institution: Rosa Zarate de Peña Headquarters	Local
		Municipal Units of Agricultural Technical Assistance (UMATA) Mayor's Office of Dagua	Local
		Corporación Autónoma Regional del Valle del Cauca (CVC) (environmental authority)	Regional
		Environmental Police	Regional
		Government of Valle del Cauca: Office of Agriculture, Fisheries and Rural Development and Office of Sustainable Development	Regional
		National Learning Service—SENA	National
		National Institute of Roads INVÍAS—Colombia	National
		Transportation and logistics company for liquid hydrocarbons in Colombia—Cenit	National

(continued)

Table 11.2 (continued)

Type of stakeholders	Description	Identified stakeholders	Scale
Academic and research	They are the education and research institutions, whether public or private, relevant to Atuncela	Javeriana University	Regional
		Icesi University	Regional
		Universidad del Valle	Regional
		National University of Colombia	National
		Alexander von Humboldt Biological Resources Research Institute	National
Non-governmental organizations (NGOs)	They are civil society organizations that develop social, environmental, productive, or cultural projects in Atuncela	Civil Defence	National
		Corporación Ambiental y Forestal del Pacifico—CORFOPAL	Regional
Private	They are companies that have a presence in the territory	Celsia energy company	Regional

Source: Authors' elaboration 2023

stakeholders have relationships with the IMRDSEA on differing scales according to the benefits that the territory provides to them. Local stakeholders maintain close and stable relationships with IMRDSEA as stewards for environmental conservation since they not only benefit from the ecosystem services that the area provides, but also act for their conservation, protection, and management.

The regional and national stakeholders are framed in relationships of interest and benefit, be it academic, investigative, management, or use of the ecosystem services provided by the area, with national stakeholders being a minority. However, notable relationships include the close relationships for conservation and management between the IMRDSEA and the environmental authority *Corporacion Autonoma Regional del Valle del Cauca* (CVC), the Mayor's Office of Dagua, the community action boards, and the civil society reserves' owners. Likewise, the area has close relationships with ASPAT, which produces panela, ASOPROCAT, an association of agricultural producers, and with some local venture companies such as the Xerofitica venture, an entrepreneurship that sells ornamental xerophytic plants, and a nature tourism venture. Regarding agricultural producers, a tense relationship was identified, since some productive activities have affected ecosystem integrity (Fig. 11.4). However, the community sugar mill enterprise has reduced the impact of productive activities, especially for the production of brown sugar.

It is worth mentioning that although the CVC, as the environmental authority, is the entity responsible for the administration of the IMRDSEA SEPL, because it is a protected area, this landscape has a participatory governance scheme that includes local actors and other stakeholders in its management. In a scheme called the Co-management Committee, consultations, agreements, and participation are fundamental factors for the governance of the area, taking into account ethno-cultural and political factors in decision-making.

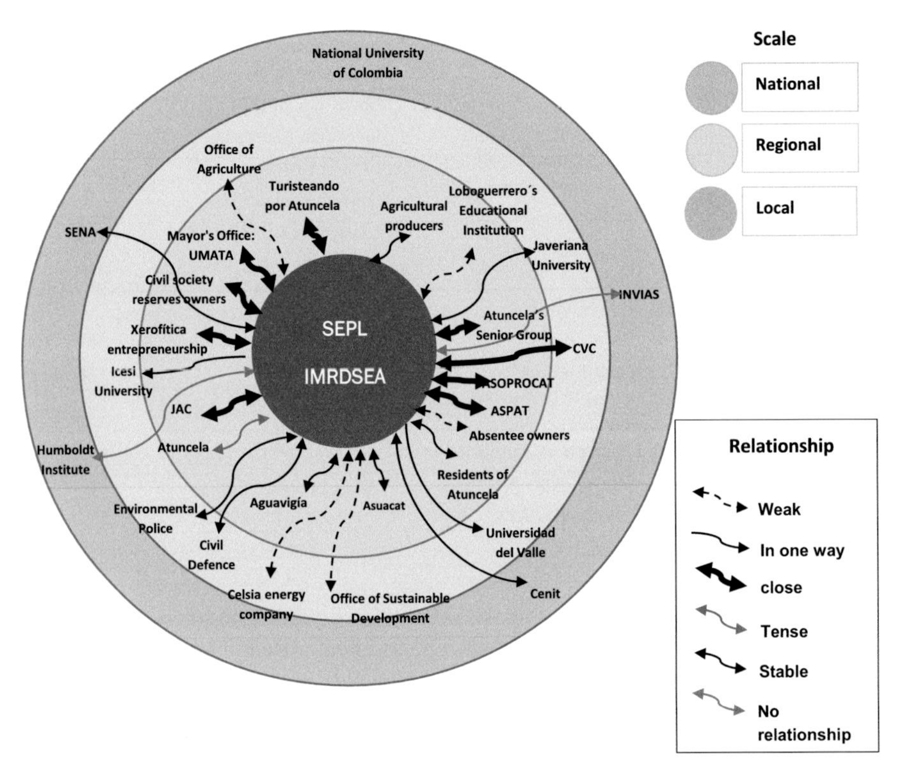

Fig. 11.4 Relationships between the stakeholders and the IMRDSEA SEPL and the scale of benefits of from the protected area. (Source: Authors' elaboration 2023)

3.2 Characterization of Productive Activities in the IMRDSEA

The inhabitants of the IMRDSEA are mostly farmers, with a subsistence economy based on medium and small productive units. Also, some inhabitants work as day labourers in large productive units. In the IMRDSEA territory, there are 120 properties, of which 36 (30%) maintain wild cover in different states of conservation. Likewise, 25 properties (20.8%) have not developed productive activities, but 82 properties (68%) are utilized for agricultural activities. Of the latter, 11 farms (13.4%) reported sugarcane as the only crop, produced by monoculture; 5 farms (6.1%) reported citrus as the only crop; 5 farms (6.1%) reported monoculture of other crops (banana, passion fruit, and tomato); and in 2 farms (2.4%) food products are only grown for food security. The remaining 59 farms (72%) engage in polyculture and of these 28 farms grow sugarcane in combination with other crops. Sugarcane is processed as panela for the market (Fig. 11.5), and other crops include fruits like lemon, mango, papaya, tangerine, guava, avocado, and orange, and produce such as coffee, cocoa, and mangosteen. Tomato, paprika, and beans are also cultivated in greenhouses for the market. Moreover bananas, cassava, squash, vegetables, and medicinal herbs are subsistence crops.

Fig. 11.5 Panela production process and finished product. (Photo: Castaño 2022)

In the IMRDSEA, 68 properties (57%) produce livestock products, including bovine, Creole hens, commercial laying hens, broiler chickens, pigs, fish, goats, and ducks. On 24 farms (33%), there is a unique livestock species: Creole chickens, goats, pigs, fish, or cattle. Furthermore, 14 farms (20%) have livestock under an extensive system, and 17 farms (25%) have pigs in intensive systems. Likewise, 48 farms (70%) have poultry, with Creole hens predominating in semi-intensive and extensive systems. A total of 24 farms (35%) have fish in intensive systems, and 7 farms (10%) raise goats in semi-intensive systems.

Agricultural practices in the IMRDSEA are carried out mainly manually, with minimal tillage and use of soil covers. Likewise, it is common to use chemical products to restore soil fertility and to control pests and diseases. With respect to water, very few properties have a concession for the use of water for agricultural activities, with most using community aqueducts or direct intakes from streams or springs. The latter generates a stressful relationship between farmers and other community aqueduct users and impacts the conservation of the IMRDSEA. Although tension exists in the IMRDSEA between productive activities and the conservation of this important ecosystem, some undertakings have reduced negative environmental impacts and others, like the community sugar mill explained below, have allowed for the use of ecosystem goods and services while promoting positive changes for the conservation of the IMRDSEA and the well-being of its inhabitants.

As described above, sugarcane cultivation has been very important for the production of panela in the study area; however, its processing has been a significant driver of environmental change, given that the 25 artisan sugar mills have usually used dry biomass from the ecosystem or even tires as fuel. However, in 2016, 25 polluting mills were replaced by a community enterprise called *Trapiche Asociativo* (Associative sugar mill), where more industrialized processing tasks are now carried out, giving better characteristics to the panela product and reducing the negative environmental impact by using bagasse from the cane as fuel (Fig. 11.6). The foregoing has allowed environmental conditions in the IMRDSEA to improve, and has also improved the livelihoods of the 48 families that are associated with the mill since they are selling higher quality organic panela. According to Luis Alfonso Tello, a legal representative of the Paneleros Association of Atuncela (ASPAT),

Fig. 11.6 View of the community sugar mill (*Trapiche Asociativo*) and bagasse from the cane used as fuel. (Photo: Castaño 2022)

"The construction of this sugar mill improved the environmental conditions. Today we have drinking water for the sugar mill, we have wastewater treatment, and we improved the panela quality, but also improved the production processes" (Arango-Muriel 2019).

These changes in productive processes in the territory have also led to the appearance of other initiatives in the territory with very low environmental impact that allow for the use of ecosystem goods and services. Tourism has been one of these economic initiatives that have gradually developed in the IMRDSEA. Taking advantage of the landscape and the particular characteristics of the subxerophytic ecosystem, e.g. the ecological adaptations of flora and fauna in the face of the limited availability of water, the community formed a group called the *Turisteando por Atuncela* that is dedicated to bird watching in the territory. Another venture born in Atuncela is Xerofítica, a virtual store for cacti, succulents, and some indoor plants. This brand seeks to publicize the subxerophytic enclave, particularly the biodiversity of the territory. It has participated in different fairs and entrepreneurship events with an environmental focus.

3.3 Evaluation of Ecosystem Services (ES) in the IMRDSEA

This study identified 23 ecosystem services associated with the four types of ecosystem services (Millennium Ecosystem Assessment 2005). The most common is the provisioning type with 11 ecosystem services (agriculture, water, sand and rock, biomass, native fruit species, wood, fibres and resins, livestock, natural ingredients, fishing and aquaculture, and medicinal plants), followed by eight regulating ecosystem services (carbon storage and capture, local/regional climate regulation, biological pest control, erosion control, soil formation, seed dispersal, water regulation, and pollination). The rest included two cultural ecosystem services (sense of belonging/traditional knowledge and tourism or aesthetic appreciation), and two supporting (conservation of genetic diversity and habitat for species).

Among the ecosystem services of the IMRDSEA, water and water regulation stand out as key to maintaining natural cover and soil configuration. These, in turn, are important for the inhabitants due to the dependency on them both for domestic use and for livestock and agriculture. Precisely, the dependency analysis identified the ecosystem services on which the community depends the most to be water, water regulation, and soil formation, these being the ones that allow residents to engage in economic activities. The lowest dependency values were obtained for timber forest products, recreation, and tourism. Reasons for the lower dependency are that community members' livelihoods are not closely related to timber forest products. Likewise, in the case of recreation and tourism, the community reports the *Turisteando* initiative as the only tourist and recreational activity, so dependence is minimal.

Despite water being identified as the most important for the communities within the IMRDSEA (cited by 100% of those surveyed), the condition of this ecosystem service is considered to be poor and it is under high pressure due to the community's dependence on it. Quantitatively, this service is in a critical state. In the first instance, it is a scarce resource in the area, and the water use index indicates high demand pressure, especially from productive activities. Moreover, the water shortage vulnerability index shows a high fragility in the face of drought events. Therefore, activities associated with water conservation are a priority due to the possibility of shortages for current human uses and negative consequences for this dry and very dry ecosystem.

Regarding trends in the above-mentioned ecosystem services, around 70% are considered to be increasing, and among these are water, erosion control, agriculture, carbon storage and capture, pollination, seed dispersal, biological pest control, medicinal plants, native fruit species, and biomass. This increase can be associated with conservation and preservation actions being carried out in the area since the declaration of the protected area and the construction of the community sugar mill. Additionally, other reported ecosystem services remained stable over time.

3.4 Relationships Between Stakeholders and the Productive Systems and Conservation of the IMRDSEA

The local stakeholders that have direct relationships and dependency on the ecosystem services are the ones that have a greater degree of involvement in conservation actions in the protected area (Fig. 11.7). Making up the majority of members of the Co-management Committee that is the governance scheme of the IMRDSEA, Aguavigía and ASUACAT are two actors directly related to the water regulating ecosystem service as they are aqueduct user associations. However, even though these stakeholders are key in the management of the protected area, they are involved in limited incidences of decision-making in the IMRDSEA, constituting a challenge for community organization.

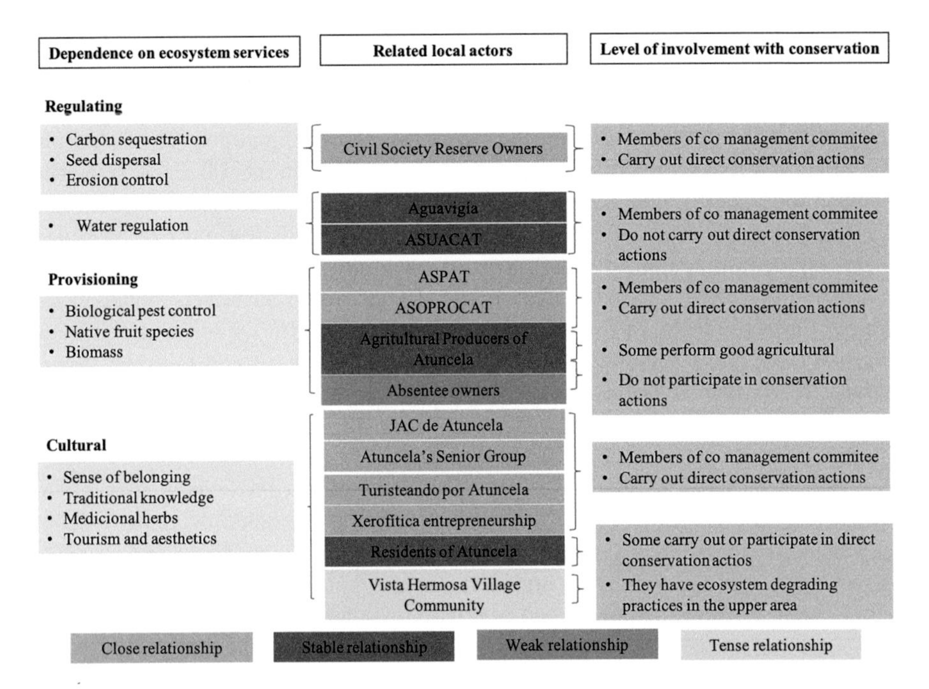

Fig. 11.7 Relationships of local stakeholders with productive systems and conservation of the IMRDSEA SEPL

Institutional stakeholders are fundamental to the protected area management and the implementation of conservation actions, although they do not directly depend on the ecosystem services (Fig. 11.8). The Mayor's Office of Dagua stands out as the territorial authority, as well as the CVC as the environmental authority. These stakeholders have led public policy and investments for conservation in the area. The study identified their participation in Atuncela to be limited to specific projects. Academic and research institutions played a role in advancing knowledge of the protected area, and from this group the Javeriana University stands out for its specific research projects in Atuncela. Likewise, companies such as Cenit and Celsia are potential actors for investment in conservation actions.

3.5 *Key Implications for Business and Biodiversity in the Context of the IMRDSEA*

The reciprocal connection between business and biodiversity is evident in the IMRDSEA SEPL, given that the inhabitants have access to natural resources, food, medicine, and drinking water, among other benefits of biodiversity, not only for their survival and well-being, but also for the development of businesses they promote for the conservation of the ecosystem. Socioe-conomic activities are carried

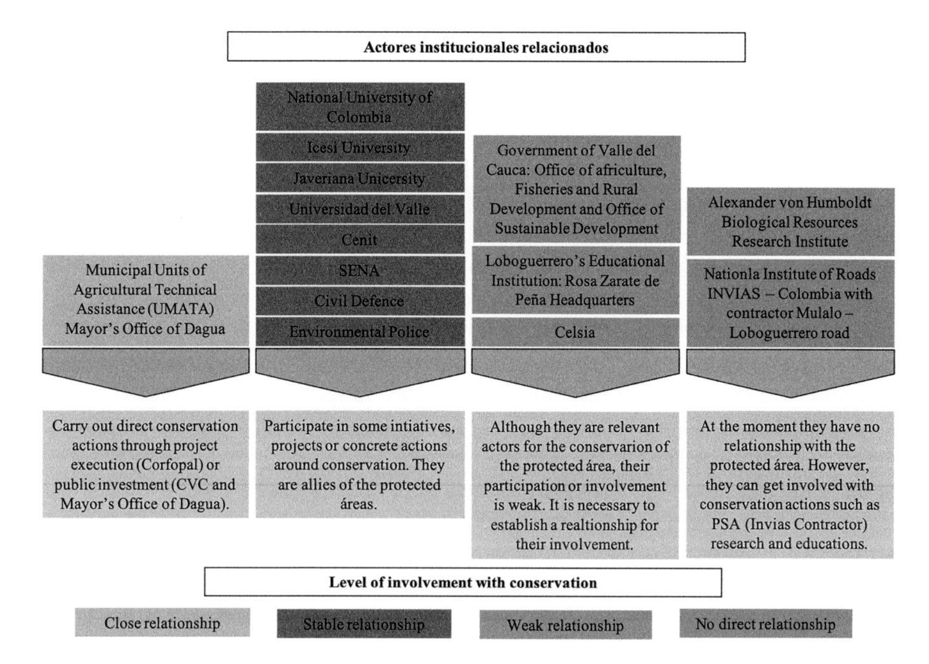

Fig. 11.8 Relationships of the institutional stakeholders with productive systems and conservation of the IMRDSEA SEPL

out on small and medium scales, such as agricultural production, livestock activities, and the commercialization of ornamental plants. Also, 73% of the 460 inhabitants of Atuncela depend economically on the large-scale cultivation of sugarcane and the production of panela (CORFOPAL and CVC 2022). These productive activities are based on the biodiversity and ecosystem services offered by the near-desert landscape of the tropical dry and very dry forest ecosystem.

Although the vast majority of small- and medium-scale productive activities within the IMRDSEA SEPL are part of the subsistence economy for food autonomy, all economic activities are based on a small-scale commercialization and a relatively medium degree of technology that allows for acceptable living standards for the landowners. In this sense, the diversity of food species produced in this landscape not only enhances human well-being, but also biodiversity, given that according to Nicholls and Altieri (2007), greater diversity in an agricultural system (as previously evidenced) entails greater diversity in the associated biota and thus favours the biodiversity of a SEPL. While complex and multispecies systems tend to have greater total productivity and capacity for regulating pests, diseases, and weeds, biodiversity improves the recycling of nutrients and energy in these productive systems (Nicholls and Altieri 2007).

Regarding large-scale businesses, the participatory and sustainable production of panela stands out because this community decided that the production of panela would no longer be done individually but in a community sugarcane mill managed by themselves. In this manner, 130 hectares of sugarcane are cultivated and processed to make panela, reducing negative environmental impacts on the SEPL and contributing to the well-being of the community. Prior to the community sugar mill, there were 25 polluting sugar mills that used tires and firewood for fuel. The inadequate production practices and low technification of these mills also resulted in low-quality panela, which was sold at a lower price (CORFOPAL and CVC 2022).

The Atuncela community sugar mill produces 25 tonnes of organic panela monthly and 280 tons per year, which represents approximately 770 million Colombian pesos (USD 183,902 based on the 5 October 2023 exchange rate of 4187 pesos to the dollar). The product is marketed in Dagua and Buenaventura and also generates 100 indirect and 15 direct jobs (Castro 2021; CORFOPAL and CVC 2022). This community initiative also has a revolving community fund, where the 48 families associated with ASPAT collect money as savings, and any of the associates can borrow to finance their productive initiatives. The amounts to be collected are set according to panela prices in the market.

In April 2022, ASPAT with the technical support of CORFOPAL, a local NGO, sent a request to the regional environmental authority for the expansion of the protected area of Atuncela. The request was accepted by the environmental authority and studies and participatory processes are being carried out to expand the area by approximately 1400 hectares. In this sense, the reciprocal relationship between the participatory sustainable production of panela business and the conservation of the tropical dry and very dry forest is evident.

4 Conclusions and Lessons Learned

The inhabitants of the IMRDSEA SEPL have recognized the importance of natural resources, not only for the development of their economic activities but also as part of their territorial identity. This favoured local stakeholder partnerships, looking forward to conservation of their natural resources, as evidenced in the participatory and sustainable production of panela, where a more technical and sustainable processing improves the panela and promotes the conservation of nature. This has positioned the community as the main supplier of this food for Dagua and Buenaventura, producing 80% of the panela consumed in these municipalities.

Thanks to the construction of the community sugar mill, there is a growing awareness among the agricultural community regarding the harmful impact of conventional agriculture, the inappropriate use of combustion sources for panela production processes, and the use of pesticides. In addition, thanks to the associative work of Atuncela's families, other benefits are evident, such as access to new sources of financing to implement transition processes towards alternative agriculture and organic production.

Some of the outcomes of this growing awareness are the incorporation of good sustainable production practices, where a considerable portion of agricultural producers take advantage of organic waste for the production of brown sugarcane (as a source of combustion). In addition to the production of fertilizers that are returned to the same crops, cyclical processes are carried out, optimizing matter and energy. Also, reducing dependence on inputs in production processes and the deterioration of the environment favours the conservation of one of the most threatened ecosystems in the world.

The remarkable community partnership in the IMRDSEA SEPL is a facilitator for the development of participatory planning and investment processes. For this reason, it is imperative to continue strengthening organizational and training processes so that the community can create and energize their own organizations and businesses based on biodiversity and the ecosystem services offered by this SEPL. Likewise, it is necessary for government institutions to have up-to-date and detailed information on the productive activities carried out in the SEPL in order to facilitate decision-making regarding management and conservation of the IMRDSEA.

Furthermore, the study of ecosystem services allowed for an exploration of the state of the IMRDSEA SEPL based on human use, finding that 23 ecosystem services were identified as having an increasing trend and permanence over time. Therefore, if activities are regulated and zoned within the IMRDSEA SEPL and new restoration and conservation activities based on associative processes are established, it may be possible to improve the quality and increase the amount of ecosystem services. Additionally, it is important to regulate the use of provisioning ecosystem services (e.g. water) to ensure usage is consistent with permitted activities and the conservation objectives of the IMRDSEA SEPL as a protected area.

In addition, it is important to pay attention to cultural ecosystem services due to the potential for ecotourism in the IMRDSEA SEPL and the presence of different links in the nature tourism value chain present in this landscape. These represent innumerable possibilities for greater articulation between the existing initiatives, but must always be carried out protecting the objects of conservation that give meaning to the IMRDSEA as a protected area. In this sense, the main objective should be the protection and conservation of the dry and very dry (subxerophytic) ecosystems. Hence, moving towards the consolidation of the tourism sector should consider restrictions in land use that may apply and positioning sustainable tourism.

Finally, it is expected that the management of the IMRDSEA SEPL advances towards the construction of an integrative vision and practice, including the local actors dependent on ecosystem services who lead conservation actions and the institutional actors who have the technical and financial capacities to get involved in the management of the protected area. This would allow for the continuous improvement of agricultural practices, natural resource use, and planning for a more sustainable management of the area. This purpose is projected in the medium term on agreements that actors have established in the governance strategy, in which each member of the Co-management Committee assumes a role in conservation. The first challenge is to improve the articulation between local community organizations and institutional actors at all levels. The second is to achieve the transition of productive activities towards more sustainable and ecological models for a territory with an agricultural vocation. The third challenge is the sustainability of the SEPL's management with a view to the conservation of ecosystem services, the preservation and protection of biodiversity and conservation objects, and the reciprocal connections between the economic and social development of the inhabitants of the territory and biodiversity.

Acknowledgements We are grateful to the community of Atuncela for their kindness and hospitality. Special thanks to the Co-management Committee who facilitated the logistics for our stay in these communities. This study was carried out within the framework of the agreement No. 187 of 2021: Combine technical efforts and economic and human resources in order to conduct participative update the management plan of the regional integrated management district of the Subxerophytic Enclave of Atuncela, located in the municipality of Dagua, department of Valle del Cauca, co-financed between the CVC and CORFOPAL. The authors are also grateful to Fernandes J., Larraniaga, V., and Campo L. for their support in fieldwork and review of this manuscript.

References

Alexander von Humboldt Biological Resources Research Institute (2015) Caracterización biológica de la ventana de biodiversidad enclave Atuncela, Municipio de Dagua, Valle del Cauca, 587 registros. Aportados por Palacio R, Muñoz A, Ospina JP, Vergara M, Sánchez GC, Echeverri C, Lizarazo J, Vargas-Méndez E, Guerrero C, Guevara-Molina C, Versión 1.1, Bogotá DC. https://doi.org/10.15472/sdkmfn [Preprint]

Alvarado-Solano DP, Otero Ospina JT (2015) Distribución Espacial Del Bosque Seco Tropical En El Valle Del Cauca, Colombia. Acta Biológica Colombiana 20(3):141–153 https://doi.org/10.15446/abc.v20n3.46703

Arango-Muriel KV (2019) Trapiche panelero les cambió la vida a 48 familias en Dagua. Noticia CVC, 9 July

Arcila-Cardona AM, Valderrama C, Chacón de Ulloa P (2012) Estado de fragmentación del bosque seco de la cuenca alta del río Cauca, Colombia. Biota Colombiana 13(2):86–100

Blackie R, Baldauf C, Gautier D, Gumbo D. Kassa H, Parthasarathy N, Paumgarten F, Sola P, Pulla S, Waeber P, Sunderland TCH (2014) Tropical Dry Forests: The State of Global Knowledge and Recommendations for Future Research. Discussion Paper. CIFOR, Bogor, Indonesia

Castro D (2021) Con recursos de Regalías por más de $1.200 millones, Gobierno del Valle benefició a 15 asociaciones productoras de panela. Informativo Gobernación del Valle, 3 January

CORFOPAL & CVC (2022) Informe técnico final del Convenio No. 187 de 2021: Aunar esfuerzos técnicos y recursos económicos y humanos con el fin de actualizar participativamente el plan de manejo del distrito regional de manejo integrado del Enclave Subxerofitico de Atuncela, localizado en el municipio de Dagua, departamento del valle del cauca. Corporación Ambiental y Forestal del Pacífico and Corporación Autónoma Regional del Valle del Cauca, Santiago de Cali

Critical Ecosystem Partnership Fund (2021) *Perfil de Ecosistema: Hotspot de Biodiversidad de los Andes Tropicales Actualización 2021*. Arlington, VA

DANE (2023) National Administrative Department of Statistics, viewed 28 January 2023. Retrieved from www.dane.gov.co

DRYFLOR et al. (2016) Plant diversity patterns in neotropical dry forests and their conservation implications. *Science* 353(6306):1383–1387. https://doi.org/10.1126/science.aaf5080

Etter A, McAlpine C, Possingham H (2008) Historical Patterns and Drivers of Landscape Change in Colombia since 1500: A Regionalized Spatial Approach. Ann Assoc Am Geogr 98(1):2–23

Etter A, van Wyngaarden W (2000) Patterns of landscape transformation in Colombia, UIT emphasis in the Andean Region. Ambio 29:432–439

García H, Corzo G, Isaacs-Cubides PJ, Etter A (2014) Distribución y Estado Actual de Los Remanentes Del Bioma de bosque seco tropical En Colombia: Insumos Para Su Gestión. In: Pizano C, García H (eds) El bosque seco tropical En Colombia. Instituto de Investigación de Recursos Biológicos, Alexander von Humboldt, Bogota, pp. 228–251

Geilfus F (2008) 80 Herramientas Para El Desarrollo Participativo. Diagnóstico, Planificación Monitoreo y Evaluación. Instituto Interamericano de Cooperación para la Agricultura (IICA), San José, C.R.

Gómez N, Reyes M, Hernandez M, Rojas Y, Salazar ML, Ayala CM (eds) (2007) Construcción colectiva del sistema departamental de áreas protegidas del Valle del Cauca (SIDAP VALLE): Propuesta Conceptual y Metodológica. Primera edición. Corporación Autónoma Regional Del Valle Del Cauca-CVC. Dirección Técnica Ambiental. Grupo Biodiversidad, Santiago de Cali

Kawulich B (2005) La observación participante como método de recolección de datos. FQS Forum: Qualitative Social Research 6(2)

Kok M, Alkemade R (2014) How Sectors Can Contribute to Sustainable Use and Conservation of Biodiversity. CBD Technical Series. PBL Netherlands Environmental Assessment Agency, The Hague

Maass J, Balvanera P, Castillo A, Daily GC, Mooney HA, Ehrlich P, Quesada M, Miranda A, Jaramillo VJ, García-Oliva F, Martínez-Yrizar A, Cotler H, López-Blanco J, Pérez-Jiménez A, Búrquez A, Tinoco C, Ceballos G, Barraza L, Ayala R, Sarukhán J (2005) Ecosystem services of tropical dry forests: insights from long-term ecological and social research on the Pacific Coast of Mexico. Ecology and Society 10(1)

MADS & ANLA (2018) Metodología general para la elaboración y presentación de estudios ambientales. Ministry of Environment and Sustainable Development and Autoridad Nacional de Licencias Ambientales, Bogotá DC

Miles L, Newton AC, DeFries RS, Ravilious C, May I, Blyth S, Kapos V, Gordon JE (2006) A global overview of the conservation status of tropical dry forests. Journal of Biogeography 33(3):491–505. https://doi.org/10.1111/j.1365-2699.2005.01424.x

Millennium Ecosystem Assessment (2005) Ecosystems and Human Well-being: Synthesis., Assessment of Climate Change in the Southwest United States: A Report Prepared for the National Climate Assessment. Island Press, Washington, DC. https://doi.org/10.5822/978-1-61091-484-0_1

Nicholls C, Altieri M (2007) Conversión agroecológica de sistemas convencionales de producción: teoría, estrategias y evaluación, Ecosistemas: Revista científica y técnica de ecología y medio ambiente 16(1):2

Orjuela-Salazar S, Gaitán MC, Quintero-Angel A (2018) Conservation on private lands integrating sustainable production and biodiversity in the Mid Dagua River Basin, Colombia. In: UNU-IAS & IGES (eds) Sustainable Use of Biodiversity in Socio-ecological Production Landscapes and Seascapes (SEPLS) and its Contribution to Effective Area-based Conservation. Satoyama Initiative Thematic Review vol. 4. United Nations University Institute for the Advanced Study of Sustainability, Tokyo, pp. 75–84

Peña-Vera T, Pirela-Morillo J (2007) La complejidad del análisis documental. *Información, Cultura y Sociedad* 16:55–81

Pizano C, García H (eds) (2014) *El Bosque Seco Tropical en Colombia*. Instituto de Investigación de Recursos Biológicos Alexander von Humboldt (IAvH), Bogotá DC, Colombia

PRATEC (Proyecto andino de tecnologias campesinas) (2012) *Dialogo de saberes. Una aproximacion epistemologica.* Biblioteca Nacional del Perú N°: 2012-05863

Rodríguez GA, Polo SM, Buitrago AM (2020) La agroindustria panelera impulsando el desarrollo rural en Colombia: un diagnóstico de la cadena productiva. Diseño e Impresión Editorial Roffaprint Editores S.A.S., Bogotá

Siyum ZG (2020) Tropical dry forest dynamics in the context of climate change: syntheses of drivers, gaps, and management perspectives. Ecological Processes 9(1):25. doi:https://doi.org/10.1186/s13717-020-00229-6

Sukhdev P, Wittmer H, Miller D (2014) La Economía de los Ecosistemas y la Biodiversidad (TEEB): desafíos y respuestas. In Helm D, Hepburn C (eds) Nature in the Balance: the Economics of Biodiversity (La naturaleza en equilibrio: la economía de la biodiversidad). Oxford University Press, Oxford
WWF (2013) Tropical and subtropical dry broadleaf forests. South America: in the Cauca Valley of western Colombia. http://www.worldwildlife.org/ecoregions/nt0207

Andrés Quintero-Ángel PhD candidate in environmental sciences and Scientific and Research Director of Corporación Ambiental y Forestal del Pacifico (CORFOPAL), Colombia. He majored in conservation and use of biodiversity with ethnic communities.

Sebastian Orjuela-Salazar Specialist in planning, follow-up and monitoring of the Fund for Environmental Action and Children - Fondo Acción.

Diana Saavedra-Zuñiga Social Worker specializing in social management, with 14 years of experience as a consultant and coordinator of disaster risk management projects, natural protected areas, and watershed management plans.

Alejandro Castaño-Astudillo Biologist specializing in environmental education, with experience in accompaniment processes and community capacity building for environmental management, nature tourism, governance, and environmental education.

María Viviana Borda Calvache Researcher at the Research group on sustainable socio-ecological systems (SSES). She has a Master's degree in conservation and use of biodiversity.

Mauricio Quintero-Ángel Has a PhD in environmental sciences and works as a Full Professor at Universidad del Valle, Colombia. Interested in research on social-ecological systems, landscape planning, and rural development.

BY NC SA

Chapter 12
From Seed to Table: A Business Case Study on Promoting a Sustainable Agri-Food System in East China

Guanqi Li, Xin Song, and Ye Shen

Abstract This case study focuses on Yuefengdao Organic Farm, an enterprise working on seed conservation and organic farming in East China, in partnership with Chuodunshan village, Kunshan Municipal Bureau of Agriculture and Rural Affairs, research institutes, and the Farmers' Seed Network. The enterprise's work is aimed at promoting the conservation and sustainable use of rice landraces and organic rice farming, which can lead to multiple socio-ecological benefits, including habitat restoration, ecotourism, climate change adaptation, and rural vitalization. The "From Seed to Table" initiative is a comprehensive approach that links community and market with a framework for a corporate biodiversity protection strategy based on spatial-temporal dimensions. Ensuring that the economic benefits of agrobiodiversity and sustainable farming are distributed equitably among stakeholders is essential. The case study highlights the importance of investing in agrobiodiversity and sustainable farming for promoting a diversified and healthy food system transformation through biodiversity conservation, utilization of ecosystem services, and rural development. Small- and medium-sized agribusinesses have a crucial role to play in achieving sustainable and equitable agricultural systems.

Keywords Biodiversity · On-farm seed conservation · Sustainable utilization · Farmers' seed network

1 Introduction

In December 2022, the Kunming-Montreal Global Biodiversity Framework (GBF) was adopted at COP15 of the United Nations Convention on Biological Diversity (CBD). The GBF has four long-term goals for 2050 and 23 targets for urgent action

G. Li (✉) · Y. Shen
Farmers' Seed Network (China), Suzhou, Jiangsu, China
e-mail: liguanqi@fsnchina.net

X. Song
Farmers' Seed Network (China), Nanning, Guangxi, China

M. Nishi et al. (eds.), *Business and Biodiversity*, Satoyama Initiative Thematic Review, https://doi.org/10.1007/978-981-97-7574-3_12

over the decade to 2030: reducing threats to biodiversity, meeting human needs through sustainable use, benefit-sharing, and on implementation and finance. Among the 23 targets, Targets 2 and 3 (also known as the 30 × 30 target) ensure and enable the restoration of 30% of all degraded ecosystems by 2030, and the conservation of at least 30% of land, water, and seas, especially areas of particular importance for biodiversity and ecosystems, and that these are effectively managed through ecologically representative, well-connected, and equitably governed systems of protected areas and other effective area-based conservation measures, recognizing Indigenous and traditional territories. In addition, Targets 10, 15, and 16 refer to the management of areas under production sectors, including agriculture, and also address the agricultural production and food consumption factors driving biodiversity loss. The production sector, especially agriculture and food consumption, are more closely tied to natural resources and biodiversity and have become drivers of biodiversity loss. This reminds us of the importance of shifting to holistic approaches that balance economic development and socio-ecological restoration for harmonious relationships between people and nature (Selinske et al. 2023).

Governments, businesses, communities, and NGOs worldwide have made efforts to promote biodiversity conservation over the past decades. While significant progress has been made by governments and scientists in achieving the first pillar of the CBD, the role of other stakeholders such as businesses in sustainable use has been inadequately addressed. Businesses are crucial in driving economic growth and development, but they also have a responsibility to ensure that their operations do not harm biodiversity and that the benefits derived from biodiversity utilization are shared equitably. The GBF recognizes that sustainable use is critical to the success of the CBD and requires the active participation and engagement of all stakeholders, including businesses, in collaboration with Indigenous peoples and local communities (IP&LC) as the custodians of biodiversity for future generations (Obura et al. 2023). Achieving sustainable use involves utilizing biodiversity resources in a way that ensures their availability for the future, and it is necessary to develop effective strategies and partnerships that incorporate the perspectives and expertise of all stakeholders.

China, like many countries, has made great efforts to achieve CBD targets based on science-oriented biodiversity conservation approaches, and these efforts are ongoing. Over the past few decades, rural communities and farming systems in China have experienced a significant decline in agrobiodiversity and the erosion of farmer seed systems as traditional crops and varieties are being replaced by modern cultivars. The Ministry of Agriculture and Rural Affairs (MOA) reported that while there were 11,590 grain crop varieties planted in China in 1956, only 3271 varieties remained in 2014, representing a loss rate of 71.8% (Xinhua News 2019). To combat this decline and loss in crop diversity and the quantity of agricultural germplasm resources, the central and local governments have invested in expanding the national gene bank system, which has the capacity to conserve 1.5 million accessions (Xinhua News 2019).

The expansion of China's ex-situ system is a significant step towards conserving agrobiodiversity. However, in-situ and on-farm conservation face significant challenges, especially in linking on-farm conservation with sustainable use of agro-biodiversity (Song et al. 2021). At the same time, there is an increasing need among consumers for healthy and safe foods, and this is one of the drivers for the state's food system transformation and green rural revitalization.

There is an urgent need to explore integrated business models linking farming and markets to address the challenges faced by rural communities and farming systems in the framework of area-based biodiversity conservation measures for achieving the 30 by 30 target. At the same time, such measures would address the objective of sustainable use and the role of businesses, enabling a balance between economic growth and socio-ecological restoration based on a holistic approach.

We borrowed and adopted a spatial-temporal framework to examine the impact of small-scale business on biodiversity conservation and sustainable utilization in East China. Specifically, we present a case study on a small agribusiness that successfully conserved and managed genetic resources through collaboration with local communities, government, and NGOs. Our findings demonstrate that businesses can play a crucial role in promoting nature-positive outcomes and achieving sustainability goals by employing the socio-ecological production landscapes and seascapes (SEPLS) approach.

The advantage of the SEPLS framework lies in its inclusion of the interaction between humans and their surrounding nature and ecosystems in the analysis, rather than solely focusing on biodiversity conservation and ecosystem service restoration. It challenges the conventional notion that views human intervention in nature, such as agriculture, as the primary driver of biodiversity loss without distinguishing between industrial agriculture and small-scale environmentally friendly farming. The SEPLS approach allows us to differentiate between different types of agriculture and various human interventions, recognizing that they have distinct roles and impacts on biodiversity conservation and utilization. This, in turn, enables us to analyse problems, causes, and solutions in a targeted and nuanced manner. In this case study, we will observe how a particular type of nature-positive business model can offer insights into the ongoing debate surrounding the balance between biodiversity conservation and farming within the contexts of land sharing and land sparing.

The Yuefengdao Organic Farm is a small-scale organic farm situated in the municipality of Kunshan of Jiangsu Province in East China (Fig. 12.1). The farm has collaborated with the neighbouring Chuodunshan village and the NGO Farmers' Seed Network (FSN) to promote in-situ conservation and sustainable use of agro-biodiversity, while also contributing to the restoration of lost biodiversity resulting from rapid industrialization and urbanization. Chuodunshan village boasts a rich agricultural heritage, with a remarkable 6000-year history of cultivating paddy rice. FSN, the leading NGO in China, has dedicated itself to actions and policy advocacy for enhancing farmers' seed systems and community-based conservation of agro-biodiversity for over a decade. This case study showcases the farm's efforts to use a SEPLS approach, combining organic farming practices with biodiversity conservation, as well as its commitment to working with local stakeholders to achieve nature-positive outcomes.

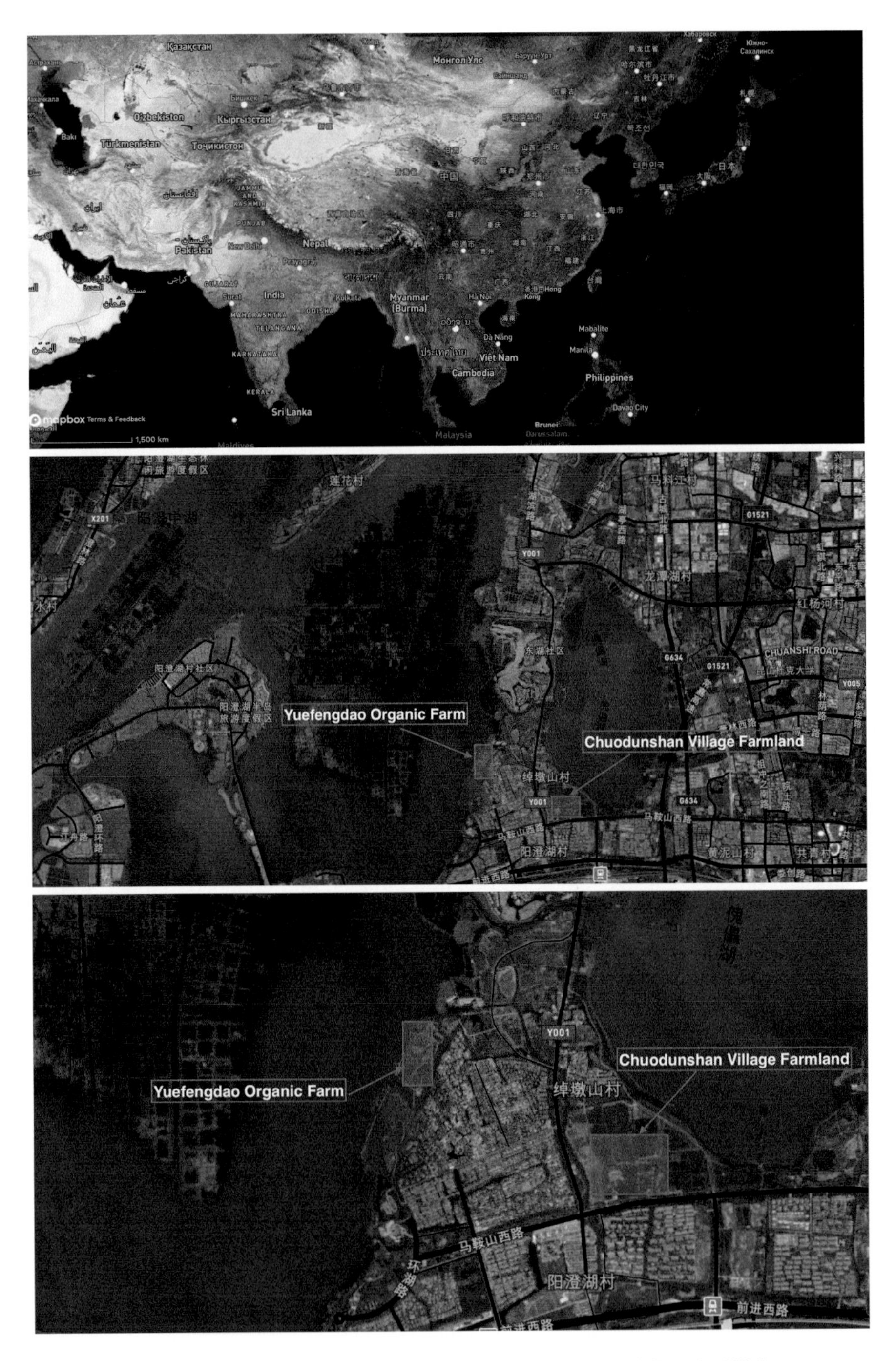

Fig. 12.1 Location of the Yuefengdao case study. (Source: Map data ©Google, 2023)

2 The Method and Activities

Biodiversity conservation and utilization can be achieved through in-situ and ex-situ conservation methods. In-situ conservation is done on-farm by IP&LC and is referred to as an informal system, while ex-situ conservation is conducted by researchers and is considered to be a formal system. Both methods have advantages and disadvantages and are often used together. In-situ conservation is crucial for maintaining genetic diversity and preserving ecosystems and species in their natural habitats, while ex-situ conservation is important for preserving endangered species. To achieve sustainable use of biodiversity, both methods are important, and the choice of method depends on the specific needs and circumstances of each species or ecosystem.

We aim to address the issue of how small-scale businesses respond to biodiversity loss and take nature-positive actions to conserve and sustainably utilize biodiversity. We were inspired by a recent study (Panwar et al. 2022) and proposed a four-cell analysis framework for business biodiversity conservation and utilization strategies (Eds. Vernooy et al. 2015, 2019) based on spatial-temporal dimensions (Fig. 12.2). The temporal dimension refers to the timing of the biodiversity

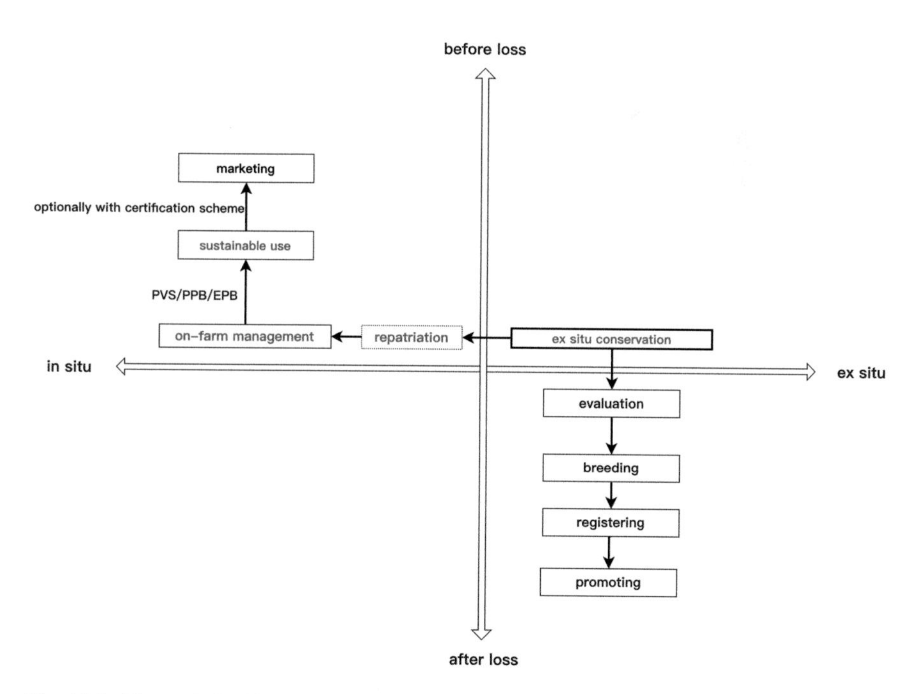

Fig. 12.2 The analytical framework for business conservation and utilization strategies. (Source: authors, based on the study)

protection initiative, which can be proactive or reactive. The spatial dimension refers to the location of the initiative, which can be in situ or ex situ.

In this analytical framework, if landraces have not yet disappeared, there are three possibilities regarding their status: they have already been collected and stored in gene banks for static preservation; they are still being cultivated by farmers; or both scenarios exist. If landraces have completely disappeared, we have no way of retrieving them. This leads to the adoption of three strategies, namely ex-situ conservation, on-farm management (one type of in-situ conservation), and repatriation.

Ex-situ conservation involves preserving landraces in gene banks without co-evolution with living circumstances and farmers' needs, where environments are strictly controlled. On-farm management entails farmers owning and cultivating landraces, utilizing these genetic resources in a living context including variable climate conditions and farmers' and consumers' changing demands. Repatriation comes into play in situations where landraces are lost in nature or are no longer in the hands of farmers. In such cases, seeds can be obtained from gene banks for the purpose of recultivation.

Combining the two dimensions of spatial-temporal intervention results in three available business biodiversity conservation and utilization strategies: ex-situ conservation, on-farm management, and repatriation. This typology provides a clearer understanding of the different strategies available for businesses' engagement and their underlying principles. It can guide businesses in developing more effective and targeted approaches to biodiversity conservation that consider the timing and location of their initiatives.

Since 2010, Yuefengdao has responded to the growing demand for healthy and safe food among emerging middle-class consumers, as well as the government's increasing focus on ecological restoration. The farm positioned itself as an advocate of organic and socially responsible agriculture, strictly managing agricultural inputs to alleviate environmental problems. In the process of practising organic farming, Yuefengdao found that while many seeds are highly commercialized and can only be obtained through seed retailers, the quality and source of the seeds cannot be guaranteed. In 2015, a baseline survey on local seed systems carried out by Yuefengdao and FSN found that most of the local landraces, including rice and beans, were disappearing, while some local vegetables were at risk of extinction.

To overcome these challenges, Yuefengdao took various actions to preserve agrobiodiversity for organic farming in collaboration with local communities and with the support of the Kunshan Municipal Bureau of Agriculture and Rural Affairs and FSN. The farm aimed to explore sustainable ways to utilize the diversity while conserving it. Activities encompassed in-situ conservation of agrobiodiversity, sustainable use of agrobiodiversity, and scaling out of the experiences gained. The following sections detail each of these three parts.

2.1 In-Situ Conservation of Agrobiodiversity

In 2015, Yuefengdao and FSN conducted a survey on traditional crop varieties in the Chuodunshan village and discovered that local landraces of rice had almost completely disappeared. The dominant rice varieties planted, such as Nangeng 46, were the result of cross breeding and promotion by agricultural research institutes. While the survey results showed that local villagers still retained some soybean and vegetable varieties, the weakening of local seed systems and farmers' seeds under the dominant influence of agricultural research institutions and seed companies was evident.

The efforts to conserve landraces and introduce new varieties have been critical in promoting agricultural biodiversity and supporting organic farming. By collaborating with Chuodunshan village, the Yuefengdao farm has been able to collect and maintain local landraces, preserving valuable genetic resources and promoting traditional farming practices. Additionally, by introducing and repatriating from gene banks, Yuefengdao has helped to diversify local crop production and improve crop resilience in the face of changing environmental and climate conditions.

The establishment of a local seed bank by Yuefengdao in 2017 has been a significant milestone in the conservation efforts. The seed bank serves as a hub for managing and conserving important crop varieties, while also providing access to seeds for organic farming and meeting consumer demand. The in-situ conservation work of the seed bank, including documentation of seed information, management of accessions, and regeneration of accessions, has been instrumental in preserving genetic diversity and promoting local food systems. With nearly 200 crop varieties currently conserved in the seed bank (Figs. 12.3, 12.4, and 12.5) and support from the local

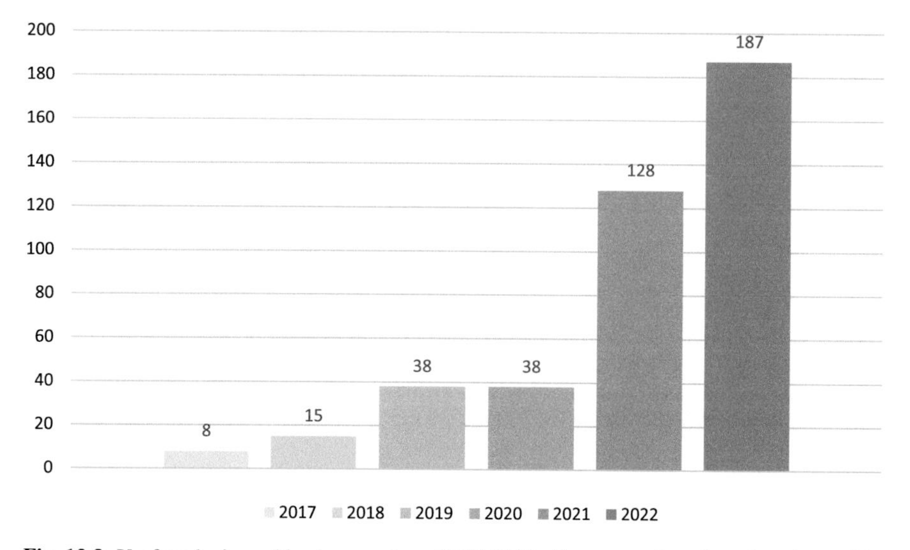

Fig. 12.3 Yuefengdao's seed bank accessions, 2017–2022. (Source: authors, based on the study)

Fig. 12.4 Yuefengdao's seed bank. (Photo taken by Zhitong Xin)

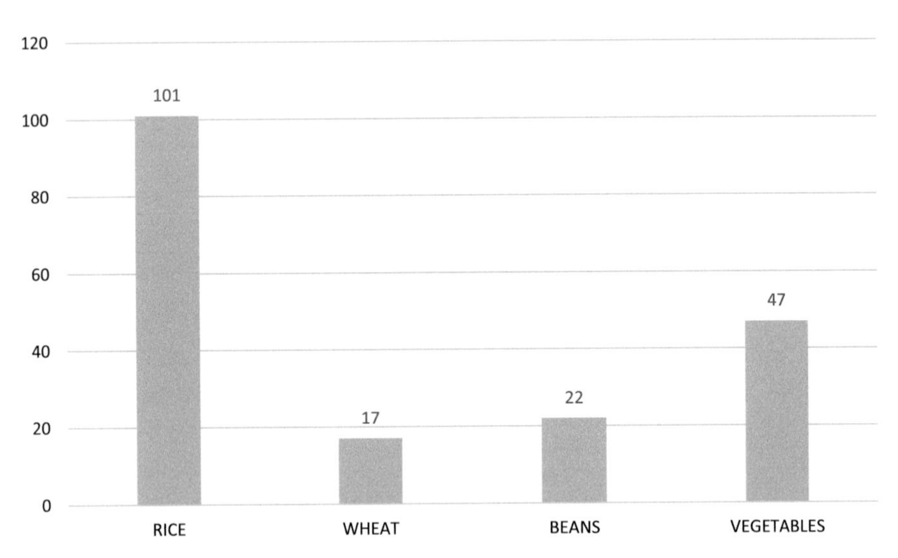

Fig. 12.5 Accessions preserved in Yuefengdao's seed bank (2022). (Source: authors, based on the study)

Fig. 12.6 Yuefengdao's demonstrative field for testing rice varieties. (Photo taken by Zhitong Xin)

government, the efforts have been crucial in promoting sustainable agriculture and ensuring food security for local communities.

2.2 *Sustainable Use of Agrobiodiversity*

To utilize the seeds conserved in the seed bank, Yuefengdao launched a "From Seed to Table" scheme to multiply the varieties. By assessing the agronomic traits and market demand of each variety in collaboration with Chuodunshan village, the scheme ensures that farmers are growing crops that are both profitable and sustainable. Additionally, by setting up parameters for height, maturation, lodging resistance, and other agronomic characteristics, the farm can guarantee the quality and consistency of the seeds, thereby improving crop yields and reducing the risk of crop failure (Fig. 12.6).

One of the most unique and prized rice varieties conserved in the seed bank is *Suyunuo*. This local sticky rice variety boasts a distinctive colour, aroma, and taste that make it highly sought after by consumers. Unfortunately, *Suyunuo* had been unadopted as a crop by local farmers and finally disappeared due to low yield, promotion of commercial varieties, and land use change. With the facilitation of FSN, Yuefengdao obtained a small amount of rice seed from the gene bank to examine the local adaptability and market response of this "lost" variety. Following 6 years of improvements and collaboration with the local community, the sowing area of *Suyunuo* has now grown to 0.4 ha as of 2022, highlighting a successful repatriated landrace.

While some qualities of *Suyunuo* make it appealing to consumers, the farm is trying to enlarge the sowing area because there is still room for improvement in terms of its resistance to lodging and pesticides. Currently, *Suyunuo* has been made into high quality rice wine and promoted in the local market. The farm established a benefit-sharing agreement with Chuodunshan village to ensure farmers can benefit from utilization of local seeds. Local agricultural extensionists have expressed interest in enhancing its agronomic and flavour traits, further demonstrating the importance of continued collaboration between public research institutes and farmers (Fig. 12.7).

The farm established a set of parameters for determining the primary rice varieties, which encompassed yield, maturity time, aroma, colour, and cooking quality. Ultimately, five varieties were chosen as the primary options for organic production, each selected based on different traits. *Bainuo* was preferred for its high yield and suitability for local food production. *Suyunuo* and *Xiangxuenuo* were distinguished by their unique aromas, even though their yields were lower, making them ideal for rice wine production. *Yaxuenuo* and *Heinuo* stood out due to their distinct colours combined with moderate yields, making them easily blendable with other ingredients for local food production. In this selection process, the farm prioritized a holistic evaluation rather than merely emphasizing high yield as the primary indicator.

The From Seed to Table scheme has achieved results—successfully balancing the farm's objectives and consumer preferences, and providing farmers with stable yield and high-quality seeds while connecting consumers with local and organic products (Table 12.1). The scheme is just one example of Yuefengdao's

Fig. 12.7 The local food made with *Suyunuo*. (Photo taken by Zhitong Xin)

Table 12.1 Basic information of the study area

Country	China
Province/State	Jiangsu
District	Suzhou City
Municipality	Kunshan Municipality
Size of the geographical area of analysis (hectare)	30
Dominant ethnicity(ies), if appropriate	Han ethnicity
Size of the case study/project area (hectare)	16.8
Dominant ethnicity(ies) of the project area, if appropriate	Han ethnicity
Number of direct beneficiaries (people)	400
Dominant ethnicity(ies), if appropriate	Han ethnicity
Number of direct beneficiaries (people)	400
Number of indirect beneficiaries (people)	2000
Geographic coordinates (latitude, longitude)	31°24'03"N 120°49'59"E

Source: authors, based on the study

commitment to sustainable agriculture and local food systems, providing a model for other organizations and communities to follow.

2.3 *Scaling Out Experiences: Spillover Effects of the SEPLS Approach*

The in-situ conservation activities and sustainable use of agrobiodiversity through organic farming highlight the vital role that gene banks and public research institutes play in supporting organic farming and conserving important crop varieties. Through the repatriation of lost seeds and the integration of traditional and modern farming practices, Yuefengdao and its partners have successfully revitalized traditional varieties like *Suyunuo*, providing a valuable resource for local farmers and consumers alike.

It has been noted that Yuefengdao employs a SEPLS approach based on conserving agrobiodiversity and utilizing this diversity for organic production in collaboration with the Chuodunshan village, generating social equity and ecological restoration outcomes. For example, Xi'an Jiaotong-Liverpool University and Yuefengdao conducted a joint study, observing one conventional and two organic farms for a period of 2 years. They developed an adaptability index to assess biodiversity conservation in organic farming and found that it increased above-ground arthropod diversity by 40% compared to conventional farming (Gong et al. 2023). Another study conducted by Xi'an Jiaotong-Liverpool University on the birds within Yufengdao farm in 2022 has revealed, in interim results shared on a social media platform, that a total of 77 bird species, comprising over 3500 individuals,

Table 12.2 Seed multiplication and production in 2022

Rice variety	Yield (kg)	Area (ha)	Seeds saved (kg)
Bainuo	13,710	1.53	80
Yaxuenuo	3360	0.86	45
Heinuo	1495	0.2	15
Suyunuo	750	0.4	30
Xiangxuenuo	460	0.13	12.5

Source: authors, based on the study

have been documented. Furthermore, researchers have identified more bird species in the transitional areas of Chuodunshan village between Yufengdao farm and lakes, revealing that these habitats may harbour greater biodiversity and be less affected by human interference (Yuefengdao 2023) (Table 12.2).

FSN has established a platform for small farmers and organic farms in eastern China to share their experiences. Through this network, Yuefengdao provides rich information and shares experience with other farmers dedicated to agrobiodiversity conservation and organic farming. In collaboration with Yuefengdao, FSN has organized workshops and a training course on agrobiodiversity conservation, including a certification scheme, to expand the reach of Yuefengdao's experiences to more farms.

3 Results

1. With support from the local government and the Farmers' Seed Network, Yuefengdao is increasing the diversity of local seeds and establishing a resilient seed system through in-situ conservation and on-farm management. This enhanced system includes a collection of local seeds, a community seed bank, and a demonstrative field for testing varieties that are open to the public. The seed bank currently holds nearly 200 types of seeds including rice, wheat, beans, and vegetables.
2. The strategy of in-situ conservation and sustainable use has spillover effects on ecological restoration. A joint study by Xi'an Jiaotong-Liverpool University and Yuefengdao observed one conventional and two organic farms in Kunshan, Jiangsu Province over 2 years. The study found that organic agriculture increased above-ground arthropod diversity by 40% compared to conventional farming.
3. Taking the five primary varieties listed in Table 12.1 as an example, in 2022, Yuefengdao achieved USD160,000 in revenue through organic production and marketing, which is five times the income achievable in conventional production. A portion of this income, totalling USD10,000, was channelled into the community as additional revenue through a profit-sharing arrangement.

4 Discussion

The issues faced by Yuefengdao reflect the decline and loss of agrobiodiversity and the weakening of the local seed system in Kunshan. The experience of Yuefengdao has demonstrated that the strategy of in-situ conservation and sustainable utilization can help preserve, acquire, utilize, and distribute seeds. To further enhance its efforts to conserve and promote agricultural diversity, Yuefengdao has been collecting and repatriating seeds in collaboration with gene banks. Through this partnership, the farm has been able to introduce new varieties and access valuable genetic resources that may have been lost otherwise. The repatriation of seeds has not only helped to diversify local crop production but has also helped to conserve important crop varieties and their associated traditional knowledge.

As agricultural policies shift towards green development and agricultural product consumption upgrades, organic agriculture offers a feasible path for revitalizing rural areas, protecting the environment, and addressing sustainable food systems and ecological conservation issues. In this context, three important points were discussed:

1. Recognizing the importance of on-farm management and sustainable use of agrobiodiversity is crucial to address the challenges faced by rural communities and farming systems in China. Preservation and use of local seed systems can achieve in-situ conservation, protecting traditional crop varieties while promoting food security and enhancing livelihoods. The formal system should focus on strengthening local seed systems and facilitating community-based seed production and distribution. Additionally, sustainable use of agrobiodiversity can promote economic development and enhance farmers' livelihoods by improving the productivity and quality of traditional varieties. Promoting sustainable farming practices and supporting the marketing of local crops and varieties can provide economic benefits to rural communities while promoting ecological sustainability.
2. Small businesses can play a vital role in promoting sustainable agricultural practices that help maintain natural habitats and biodiversity. For instance, small-scale agribusiness working on organic agriculture can utilize circular methods that combine crops and livestock to provide various benefits such as soil conservation, carbon sequestration, and habitat preservation. Support for local communities is another essential aspect, as small businesses can provide economic opportunities for people who rely on agrobiodiversity. Small-scale processors may use locally sourced ingredients to create value-added products, generating income and supporting local economies. Furthermore, small businesses can help develop niche markets for agrobiodiversity products, incentivizing farmers to maintain traditional varieties and practices. For example, small-scale specialty food producers may source unique ingredients from local farmers and promote their products to consumers who value biodiversity and traditional foods.

3. Discussing the relationship between businesses and biodiversity is crucial for promoting sustainable development and economic growth. Encouraging businesses to prioritize biodiversity conservation and sustainable practices can significantly benefit the environment and local communities. Promoting dialogue on this topic can raise awareness and identify strategies for balancing economic development and conservation efforts to ensure a sustainable future for all. Establishing a mechanism for conserving, selecting, and breeding organic seeds adapted to the local climate and ecological conditions can create a vibrant seed system, encouraging organic producers and local communities to play a greater role in the market value chain and improve the overall ecological environment. By working with local governments, gene banks, and NGOs, businesses can access a wide range of genetic resources to improve crop resilience and adaptability to changing environmental conditions.

5 Conclusion

The SEPLS approach employed by Yuefengdao serves as an example of how small-scale businesses can contribute to biodiversity conservation and utilization. The four-cell analytical framework provides a useful guide for businesses seeking to develop effective and targeted approaches to biodiversity conservation. By considering the timing and location of their initiatives, businesses can adopt a more strategic approach to biodiversity conservation and utilization, thus promoting sustainable agriculture and supporting local communities.

The efforts of small businesses like Yuefengdao demonstrate the significance of in-situ conservation and sustainable utilization of genetic resources in promoting sustainable agriculture and food security. Collaborating with local communities and organizations has allowed the farm to conserve and reintroduce valuable crop varieties, promote traditional farming practices, and support local food systems. Establishing a local seed bank has been a significant milestone in preserving genetic diversity, providing access to seeds for organic farming, meeting consumer demand, and improving crop resilience in the face of changing environmental and climate conditions.

In conclusion, small businesses can play an essential role in the in-situ conservation of agrobiodiversity, particularly in rural areas where many small-scale farmers rely on biodiversity for their livelihoods. By implementing these and other strategies, we can encourage more small businesses to take action to conserve biodiversity, promoting sustainable development and helping to ensure a healthy future for all.

References

Gong S, Zhou X, Zhu X, Huo J, Faghihinia M, Li B, Zou Y (2023) Organic rice cultivation enhances the diversity of above-ground arthropods but not below-ground soil eukaryotes. Agric Ecosyst Environ 347:108390. https://doi.org/10.1016/j.agee.2023.108390

Obura D, Agrawal A, DeClerck F, Donaldson J, Dziba L, Emery MR, Friedman K, Fromentin JM, Garibaldi LA, Mulongoy J, Navarrete-Frias C, Reidl PM, Roe D, Timoshyna A (2023) Prioritising sustainable use in the Kunming-Montreal global biodiversity framework. PLOS Sustain Transform 2(1):e0000041. https://doi.org/10.1371/journal.pstr.0000041

Panwar R, Ober H, Pinkse J (2022) The uncomfortable relationship between business and biodiversity: Advancing research on business strategies for biodiversity protection. Bus Strategy Environ 32(5):1–13. https://doi.org/10.1002/bse.3139

Selinske M, Bekessy SA, Wintle BA, Garrard GE (2023) Biodiversity needs both land sharing and land sparing. Nature 620(7975):727. https://doi.org/10.1038/d41586-023-02631-4

Song X, Li G, Vernooy R, Song Y (2021) Community Seed Banks in China: Achievements, Challenges and Prospects. Front Sustainable Food Syst 5. https://doi.org/10.3389/fsufs.2021.630400

Vernooy R, Bessette G, Otieno G (eds) (2019) Resilient Seed Systems: Handbook, 2nd Edn. Bioversity International, Rome. Available online at: https://hdl.handle.net/10568/103498

Vernooy R, Shrestha P, Sthapit B (eds) (2015) Community Seed Banks: Origins, Evolution and Prospects. Routledge, London. https://doi.org/10.4324/9781315886329

Xinhua News (2019) National Bank of Crop Germplasm Resources to be built in March with capacity of 1.5 million accessions, viewed 16 November 2020. Retrieved from http://www.xinhuanet.com/politics/2019-01/17/c_1124000762.html

Yuefengdao (2023) WeChat public accounts. New research on birds and other species in Yuefengdao organic farm. https://mp.weixin.qq.com/s/SD0JauKuizG-ZK36qXH9UA

Guanqi Li Director of East Office of Farmers' Seed Network (China). Works on projects focused on farmers' seed systems enhancement to build practice-science-policy linkage and promote agricultural biodiversity conservation and utilization that enhances food security and sustainable food systems.

Xin Song Director of Southwest Office of Farmers' Seed Network (China). Focuses on community-based participatory plant breeding and related action research to improve small farmers' resilience in Southwest China.

Ye Shen Coordinator at Farmers' Seed Network (China). Work on projects focused on seed conservation and utilization in the context of climate change to build "From Seed to Table" initiative that promotes sustainable seed and food systems.

BY NC SA

Chapter 13
Socio-Ecological and Socio-Economic Assessment of Complex Rice Systems: A Case Study in Lamongan District, Indonesia

Uma Khumairoh, Rochmatin Agustina, Euis Elih Nurlaelih, Dewi Ratih Rizki Damaiyanti, Adi Setiawan, and Jeroen C. J. Groot

Abstract Traditionally, rice agroecosystems with complex wetland and dryland habitats have a great wealth of biological and cultural diversity. After the Green Revolution, rice production underwent a dramatic transformation driven by the expansion of monoculture rice fields, high-intensity cropping cycles, and utilization of commercial rice varieties and agrochemicals. The resulting loss of biocultural diversity has amplified an urgent call for another transformation to restore agrobiodiversity. Despite numerous interventions for knowledge dissemination on agrobiodiversity, little is known about the long-term impacts of those interventions on farmers' practices and perceptions. Focusing on farm income and farmers' perceptions towards rice farming systems, we report on findings from exploratory surveys conducted 5 years after a Farmer Field School (FFS) intervention that set up complex rice system (CRS) demonstration farms in Lamongan, Indonesia. Findings showed that the highest income was achieved by farmers who cultivated more than two species on their rice bunds, followed by farmers who grew two species, with the

U. Khumairoh (✉)
Integrated Organic Farming Systems Research Centre, Faculty of Agriculture, Brawijaya University, Malang, East Java, Indonesia

Farming Systems Ecology, Wageningen University and Research, Wageningen, The Netherlands
e-mail: uma.kh@ub.ac.id; uma.khumairoh@wur.nl

R. Agustina
Integrated Organic Farming Systems Research Centre, Faculty of Agriculture, Brawijaya University, Malang, East Java, Indonesia

Faculty of Agriculture, University of Muhammadiyah Gresik, Gresik, East Java, Indonesia

E. E. Nurlaelih · D. R. R. Damaiyanti · A. Setiawan
Integrated Organic Farming Systems Research Centre, Faculty of Agriculture, Brawijaya University, Malang, East Java, Indonesia

J. C. J. Groot
Farming Systems Ecology, Wageningen University and Research, Wageningen, The Netherlands

M. Nishi et al. (eds.), *Business and Biodiversity*, Satoyama Initiative Thematic Review, https://doi.org/10.1007/978-981-97-7574-3_13

least income achieved by farmers who grew only rice. This study also demonstrated that proper selection of crops complementary to rice was a key to success, along with a slight widening of the rice bunds. Focus group discussions with three groups of farmers suggested that inputs and income were the most emphasized components of monoculture, while planting vegetables and other crops on rice bunds was an additional central component in rice farming systems with two and more than two species. These findings support the argument that diversifying rice fields could improve a rice farm's economic performance. Therefore, incorporating economic considerations into the promotion of agrobiodiversity is critical to protect and sustain rice farming businesses.

Keywords Farm income · Rice farming systems · Species selection · Farmer perception · Redesigning farms · Rice bunds

1 Introduction

The diverse composition and interaction of plant and animal species in rice agroecosystems sustains biodiversity, local food systems, and cultural ecosystem services. Accordingly, rice agroecosystems are associated with rich biodiversity. The unique habitats they provide, which include both wet fields and the relatively drier part of rice bunds, offer temporary and permanent refuge to many living organisms (Cruz-Garcia et al. 2016).

Alongside the Green Revolution, that was introduced in Indonesia from the late of 1960s (Darmawan et al. 2006), rice agroecosystems underwent dramatic transformations driven by the expansion of monoculture rice fields, higher-intensity cropping cycles, the use of commercial rice varieties, and the application of agrochemicals. These changes have negatively impacted not only floral and faunal biodiversity, but also the livelihoods of many local communities. At present, the centrality of rice farming in the agricultural and social systems of Indonesia and many other Asian countries contrasts with the significant erosion of rice-associated biological and cultural diversity in native communities. This situation has generated calls for increased attention to the links between traditional rice production systems and Indigenous rice-based customs to restore biocultural diversity and agrobiodiversity (Pfeiffer et al. 2005).

Conservation of biocultural diversity is important for a functioning ecosystem and delivery of ecosystem services. However, redesigning biodiversity-rich agroecosystems to realize these needs rests on comprehensive approaches involving technical, managerial, and educational interventions that require both short- and long-term investment. The redesign processes must also consider economic sustainability. Biodiversity conservation should eventually lead to profitable food production and the achievement of other socio-cultural goals (Feledyn-Szewczyk et al. 2016).

This chapter presents the results of an assessment of the impacts of rice farm diversification attempts, focusing on the crop diversity of rice bunds. Usually left bare, rice bunds can serve as part of a transition process from monoculture rice production to a complex rice system (CRS). The intervention was based on the Farmer Field School (FFS) method, with 25 farmers in the village of Wajik in the Lamongan District of East Java Province, Indonesia, participating in the study. The FFS method allowed for knowledge sharing among researchers, farmers, and all participants and the incorporation of participant feedback into the design of the demonstration farm. The aim of this study, therefore, was to assess the impacts of the CRS-FFS intervention implemented in 2016 on the socio-economic and socio-ecological conditions of the community in the targeted area.

2 Methodology

2.1 Site Description and Current Situation

The study was conducted from December 2020 to July 2021 in the village of Wajik, Lamongan District in East Java Province of Indonesia (Table 13.1, Fig. 13.1). The majority of inhabitants are rice farmers who grow two crops of rice a year. The altitude of Wajik is 8 m above sea level, with a mean temperature of 26 °C. The site receives an annual precipitation of 1600 mm. Soil is classified as heavy clay Vertisol. Rice bunds are created at a minimum of 1 m in width and depth to make them strong and keep them from collapsing due to water pressure, especially during heavy rains. Despite the fact that rice bunds in the study area are wider in comparison to bunds in other districts of East Java, many farmers leave them bare without a cover of crops or cash crops (Fig. 13.2).

Table 13.1 Basic information of the study area

Country	Indonesia
Province	East Java
District	Lamongan
Municipality	Lamongan
Size of geographical area (hectare)	181,000
Dominant ethnicity(ies), if appropriate	Javanese
Size of case study/project area (hectare)	281 ha with 185 ha rice fields
Dominant ethnicity in the project area	Javanese
Number of direct beneficiaries (people)	300
Number of indirect beneficiaries (people)	1,356,027
Geographic coordinates (latitude, longitude)	7°06′11″S, 112°22′32″E

Source: Statistic Indonesia, 2023

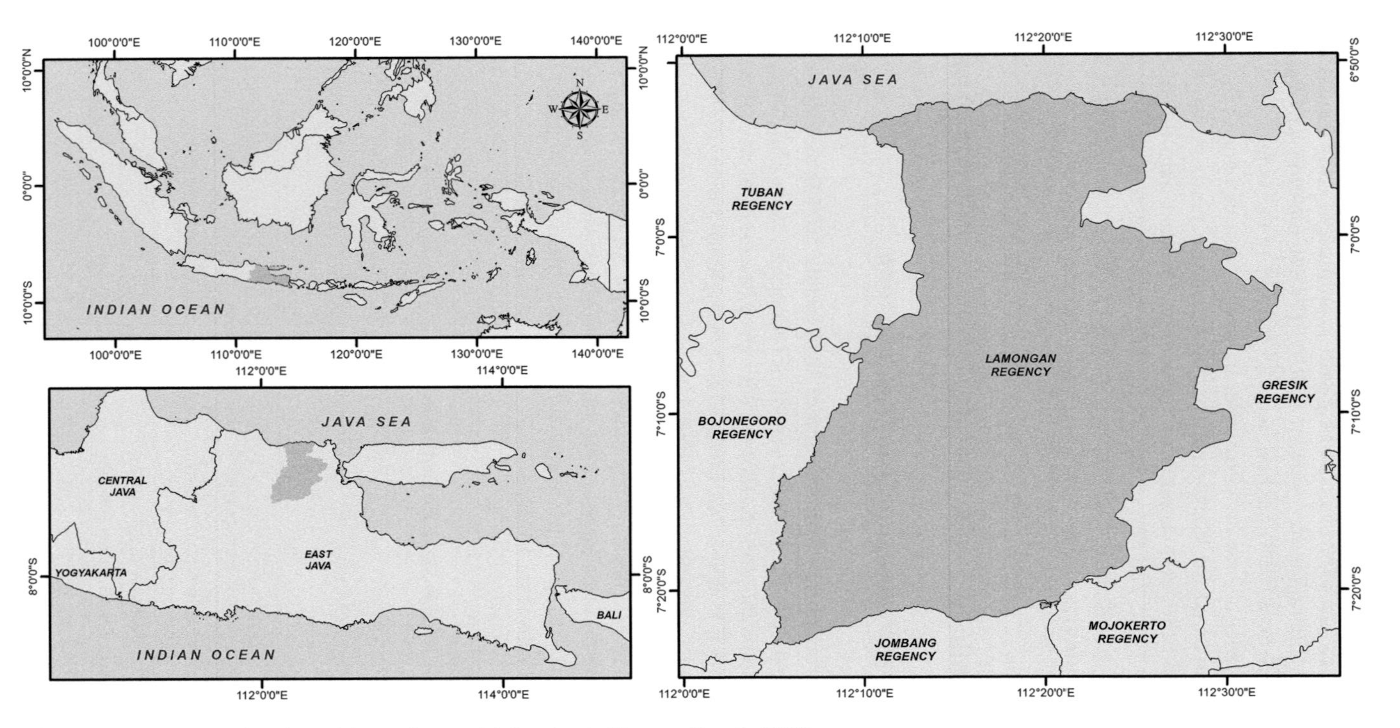

Fig. 13.1 Map showing the location of the study area (pink colour). (Source: Google 2023)

Fig. 13.2 Monoculture rice systems in Wajik, Lamongan, East Java, Indonesia. (Photo credits: Uma Khumairoh and Rochmatin Agustina)

2.2 Research Methods

CRSs in Lamongan were developed through two cycles of FFS interventions. The field schools were run from June 2014 to June 2016 and involved participatory experiments on three rice farms that served as demonstration or model farms (Khumairoh et al. 2019). The choice of plant species to be integrated was crucial for achieving the successful outcomes, which were mainly income generation and input reduction. This study aimed to evaluate the impacts of the FFS interventions while assessing the possibility of scaling up CRSs at the landscape level.

The assessment was made through surveys employing household interviews, farm visits, and a focus group discussion (FGD) approach. The FGDs were organized to assess farmers' perceptions towards their rice farming systems. A total of 55 farmers were interviewed, in either their rice fields or their homes, with questions focusing on the socio-economic and ecological aspects of their farming systems. Respondents were selected based on farm complexity or crop diversity, including (a) monoculture rice systems, (b) two-species rice farming systems, and (c) more than two-species rice farming systems. The collected data were analyzed using both descriptive and inferential statistics. ANOVA analysis was followed by a post hoc test, aiming to analyze the impact of complexity on the socio-economic and socio-ecological performance of the compared farming systems. The descriptive statistics procedure and ANOVA tests were performed in the SPSS 23 software package (SPSS Inc., Chicago, Illinois, USA).

Three FGDs were conducted with the three different farmer groups (i.e. monoculture, two-species, and more than two-species farming systems), consisting of 5–8 farmers per group, to assess farmers' perceptions of their rice farming systems. Farmers were asked to identify important components (Fig. 13.4) in their rice

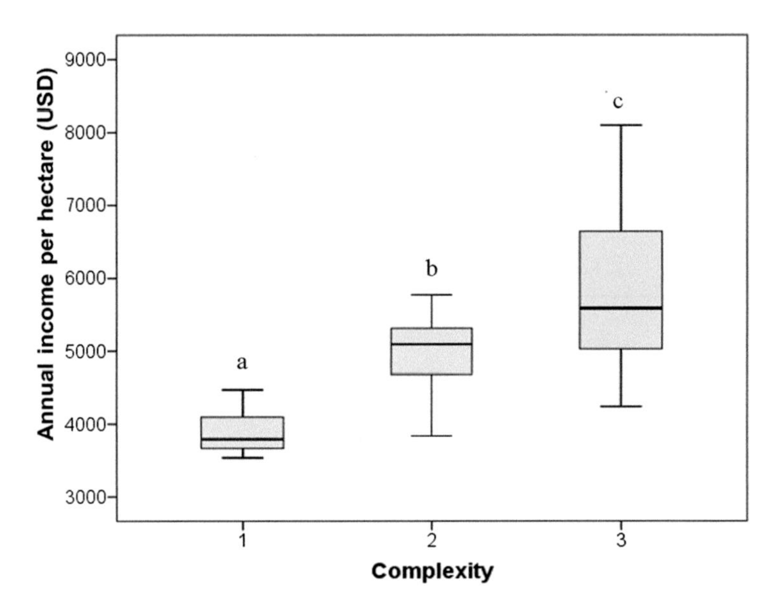

Fig. 13.4 Box plot illustrating income generation of different rice farming systems: 1, monoculture rice systems (n = 19 farm households); 2, two-species rice farming systems (n = 21 farm households); and 3, more than two-species rice farming systems (n = 15 farm households). (Source: Prepared by authors)

agroecosystems. They were then asked to draw lines among components to show the relationships among them. In the final portion of FGDs, farmers in each group were asked to rank each component in order of importance, with one being the most important and subsequent numbers indicating decreasing levels of importance.

3 Results and Discussion

3.1 *Socio-Economic Impacts*

Of the farms studied in this case, some farmers chose to grow vegetables on the rice bunds due to their higher yields and prices (Fig. 13.3), while others grew corn and legumes to support resilient livelihoods in the event of water shortages or unpredicted drought.

Survey results on the impacts of interventions indicated that the use of a participatory learning method in two cycles of FFSs contributed to improved socio-economic outcomes over the course of redesign processes into CRS (Fig. 13.4). The most complex rice systems (more than two species) had a significantly better farm income performance ($F (2, 52) = 26.468$; $P = 0.001$) due to wider rice bunds (Table 13.3) to grow more vegetables, followed by two-species rice systems. Monoculture rice production resulted in the least income for farmers despite high

Fig. 13.3 Complex rice systems in Wajik, Lamongan, East Java, Indonesia. (Photo credit: Rochmatin Agustina)

Table 13.3 Ecological and economic conditions of different rice farming systems

Rice system complexity	% of non-rice area	Non-rice area utilization	Efficiency: Profit per USD of bought fertilizer	% of farms burning straw
1 (monoculture)	12 ± 1	Bare	43 ± 2.5	54 ± 0.6
2 (two-species rice farming systems)	12 ± 0.4	Crop diversification	66.5 ± 10	43 ± 0.5
3 (more than two-species rice farming systems)	14 ± 0.5	Crop diversification	73 ± 4.5	15 ± 0.4

Source: Prepared by authors generated from survey results

rice yields because rice bunds were left bare, even though bund width was similar to that of the two-species rice farming system.

These results suggest that efforts to conserve agrobiodiversity must take economic sustainability into consideration. Merely diversifying agriculture for ecological purposes is not enough and may result in less crop yield for farmers, leading to income reduction. Programmes must also ensure that farmers can make ends meet for their households, nutritionally, economically, socially, and culturally. Appropriate species selection when increasing complexity is crucial for income generation and risk reduction.

Farmers' perceptions of their rice farming systems are illustrated in Fig. 13.5, shown by the number of components and the interactions that farmers identified. Figure 13.5a shows that monoculture farmers mentioned very few components, highlighting pesticide and fertilizer as keys to successfully producing rice that generates income. Figure 13.5b, c depict the perceptions of farmers who grew two and more than two crop species on their farms, respectively. They shared a similar number of components identified, but a different number of interactions. Despite producing two agricultural species, the concerns of farmers cultivating two crop species (Fig. 13.5b) were similar to monoculture farmers (Fig. 13.5a), emphasising the importance of chemical inputs and income. In their perception, vegetables on rice bunds also needed fertilizers, pesticides, and herbicides, showing a lack of knowledge on diverse plants that could invite natural enemies to deter pests. Meanwhile, farmers who grew more than two agricultural species (Fig. 13.5c) identified more relationships among important components. These positive effects included attracting natural enemies of pest and attracting diseases and weeds by vegetables and other border plants. These impacts led to higher rice yields and diversified farm income.

At the end of FGDs, farmers in each group agreed on the ranking by importance of components on rice production systems and farm income that they had mentioned in previous FGD sessions in order to validate and highlight each group's perceptions. These rankings are presented in Table 13.2 and show the differing levels of importance of the same components in the three rice farming system groups. Monoculture rice farmers mentioned seven important components, ranking rice production first, followed by income generation and the use of pesticides and fertilizers to make up the four most important components. Farmers in two-species rice farming systems mentioned eight important components, but listed rice production, labour input, income generation, and food provision as the four most important

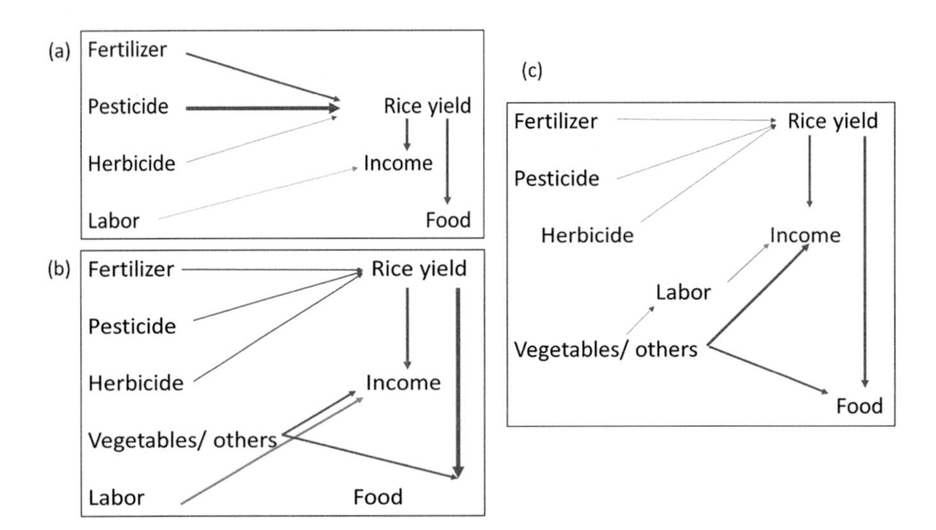

Fig. 13.5 Important components of farming systems and their relationships as identified by farmers from three different farming systems: (**a**) monoculture rice systems, (**b**) two-species rice farming systems, and (**c**) more than two-species rice farming systems. Blue and brown arrows indicate positive and negative relationships, respectively. The thickness of the arrows corresponds to the strength of the relationships, with the thickest arrows showing the strongest relationships/perceptions. (Source: Prepared by authors)

Table 13.2 Ranking by importance of components in three different rice farming systems, 1 = most important

Component	1 species (monoculture) rice farming systems	2 species rice farming systems	>2 species rice farming systems
Rice	(1)	(1)	(2)
Pesticide	(3)	(7)	(6)
Fertilizer	(4)	(5)	(5)
Herbicide	(6)	(6)	(6)
Vegetable/corn	–	(8)	(1)
Food	(7)	(4)	(3)
Income	(2)	(3)	(4)
Labour	(5)	(2)	(7)

Source: Prepared by authors

components in the order of importance. Meanwhile, despite sharing similar components, farmers in more than two species systems ranked vegetables and other crop production first, followed by rice production, food provision, and income generation as the four most important components in their farming system's activities. Knowledge and their participation in FFS-CRS might have affected these different perceptions.

3.2 *Ecological and Economic Impacts*

Table 13.3 shows rice bund dimensions and utilization that are associated with farm profits, ecological consequences of fertilizer use efficiency, and treatment to the rice straw.

Despite the similar dimensions of rice bunds in both rice monoculture and two-species polyculture, the rice bunds in monoculture systems are left bare, while those in rice polyculture systems are used to grow other crops. These results suggest that allocating slightly wider rice bunds for growing vegetables could increase production with similar costs for agrochemicals, increasing the efficiency of land and fertilizer use in the most complex systems.

Farmers' attitudes towards complexity in this study were related to their attitudes toward the environment, as indicated by responses to a survey on rice straw burning. Only 15% of farmers in complex systems, compared to more than 50% of farmers in monoculture systems, burned their rice straw. Farmers with more complex elements (species) recycled straw in their fields, either using it as mulch for crops on rice bunds or letting it decay on the rice field to become fish feed and organic fertilizer. Organic fertilizers generated from decomposed and then mineralized straw improve soil organic matter and soil carbon (55–60%), contributing to the mitigation of climate change (Romasanta et al. 2017; Amelung et al. 2020; He et al. 2023). Last but not least, flowering crops on rice bunds are a potential habitat for many fauna species that might play important roles as natural enemies of pests (Beaumelle et al. 2021). These protect agricultural products from their enemies and reduce pesticide use, environmental pollution, and production costs.

Finally, the case study results highlight the importance of utilising the space on rice bunds and the effectiveness of crop diversification in ecosystem management, increasing farm performance ecologically, culturally, and economically.

3.3 CRS Contribution to Cultural-Biodiversity Conservation and the Roles of Traditional and Local Knowledge

CRSs revive Indigenous knowledge, including integration of the following:

1. **Crops on rice bunds**
   Traditionally, Javanese farmers grew taro and legumes on rice bunds. During the Green Revolution, farming became increasingly specialized, suggesting rice bunds should be as narrow as possible and cleared of all weeds and other plants to prevent rats. On the contrary, CRSs encourage crop diversification on rice bunds and the use of other approaches for rat control, for example by growing poisonous cassava (*Manihot esculenta* Crantz). In other words, CRSs restore traditional culture and knowledge on growing plants on rice bunds, improving farmer household nutrition, adding farm income, and increasing ecosystem services for spiritual values (e.g. plants grown on rice bunds to be harvested for offerings or taken to visit graves).
2. ***Surjan*** systems, with "*Surjan*" (in Javanese) meaning "strip" (in English), is a traditional Javanese rice system where rice plots have wide bunds to grow other crops alongside rice, creating a farm design referred to as "strip cropping." Intensification of rice farming has led to rice bunds that are reduced to a maximum of 30 cm, making it almost impossible to adopt approaches that promote agrobiodiversity. The CRSs adopted "modified" *Surjan* systems, enabling the restoration of agrobiodiversity while maintaining acceptable rice bund dimensions with various designs and management.
3. ***Jajar legowo*** is a local Javanese cultivation method that involves leaving one row of rice empty to allow sunlight to penetrate to more parts of the rice plants to maximize photosynthesis. Accordingly, plant growth and yield are considered to be better than methods with less light penetration.
4. **Other examples of local knowledge** include (1) growing papaya or *Acorus calamus* as natural antiseptics and medicinal plants for human and animals (e.g. fish and ducks), (2) growing tubers on rice bunds to protect rice from rats, (3) growing *Acorus calamus* along inlet irrigation to repel pests trying to enter to rice fields, and many more.

As suggested from the above results, the agricultural sector holds a wealth of biodiversity-relevant knowledge and should be a critical player in contributing to effective ecosystem and biodiversity management. The CRS is not only an assemblage of plant and animal species, but also a combination of traditional knowledge and modern sciences (Khumairoh et al. 2019). Combining modern science with traditional knowledge could validate and explain the reasons underlying traditional practices to produce good agroecosystem-management outcomes to reconcile environmental and agricultural production goals.

3.4 Challenges in Scaling up CRSs

1. **Skills and knowledge**: With farmers increasingly specialising due to modern and industrial farming, rice farmers have lost their skills and knowledge on how to combine different plants in their rice systems. The loss of this Indigenous knowledge poses a challenge for implementation of CRSs. Therefore, investment in training farmers is required to improve their agrobiodiversity management capacity.
2. **Initial capital**: Integrating CRS components into rice systems could prove to be easy or difficult, depending on farmers' skills and the conditions at the starting point, both of which have consequences for the inputs and facilities needed for redesigning farms. Farmers who have already practised organic systems and polyculture (species diversity in the spatial dimension) or rotation (species diversity in the temporal dimension) may find it easier and cheaper to transform their farming systems from simpler to more complex systems than conventional and monoculture farmers. Conventional and monoculture farmers who want to turn their farming systems into complex systems could face challenges with high input costs and uncertain success if they lack skills and knowledge. One solution may be to include a pre-payment scheme whereby organizations and individual consumers pay for products in advance to help smallholder farmers buy inputs to start CRSs.
3. **Market and collective management**: A potential strategy to scale up CRSs is integration of elements that are marketable. However, since the majority of rice farmers are smallholders, the small production volume may not be desirable in the market, constraining the continuity of business. Thus, a landscape approach that integrates groups of diverse farmers, stakeholders, policies, and practices should be implemented to provide sufficient volume and continuity for the market and to promote the sustainability of CRS implementation.
4. **Awareness**: Consumer awareness toward ecological farming is an essential element for farmers and marketing agencies, including local sellers and wholesalers for regional markets, to successfully plan production that can capture a greater market share. This greater market share can encourage farmers to adopt CRS as a scaling-up strategy. However, consumer knowledge and awareness toward ecological farming are still very low (Suharjo et al. 2016), limiting the market absorption rate of sustainable products produced in CRSs, thus presenting a challenge to scaling up.

4 Conclusion

Our study has highlighted the importance of economic considerations in redesigning agricultural systems toward complex agroecosystems to promote agrobiodiversity and conservation. Taking economic aspects into account, biodiversity conservation could be implemented through the appropriate selection of grown and reared species with cultural and spiritual values recollected from Indigenous knowledge that supports local food systems and aligns to markets. The study suggests that increasing the width of rice bunds as a non-rice area to grow vegetables and other crops is an option that allows more flexibility in accommodating more marketable species. However, this is not necessarily the only solution when suitable species are available and preferred by farmers to be cultivated within rice plants, such as water plants and water animals. The positive perceptions of farmers towards CRSs in this study stemmed from greater economic farm performance based on the cultivation of marketable and consumable diverse species in rice systems. Despite the many benefits of CRS, scaling up CRS faces challenges related to farmer knowledge, consumer and community awareness, initial capital, and market share. Nonetheless, some solutions are available with proper training and collective management.

Acknowledgements This study has been possible with financial support from the Nestlé Foundation for the Study of Problems of Nutrition in the World, Lausanne, Switzerland and EQUITY programme 2023–2024, Brawijaya University, Malang, Indonesia.

References

Amelung W, Bossio D, de Vries W, Kögel-Knabner I, Lehmann J, Amundson R, Bol R, Collins C, Lal R, Leifeld J, Minasny B, Pan G, Paustian K, Rumpel C, Sanderman J, van Groenigen JW, Mooney S, Wander M, Chabbi A (2020) Towards a global-scale soil climate mitigation strategy. *Nat Commun* 11:5427. https://doi.org/10.1038/s41467-020-18887-7.

Beaumelle L, Auriol A, Grasset M, Pavy A, Thiéry D, Rusch A (2021) Benefits of increased cover crop diversity for predators and biological pest control depend on the landscape context. Ecol. solut. evid. 2:e12086. https://doi.org/10.1002/2688-8319.12086.

Cruz-Garcia GS, Struik PC, Johnson DE (2016) Wild harvest: distribution and diversity of wild food plants in rice ecosystems of Northeast Thailand, NJAS - Wagening. J. Life Sci. 78:1–11. https://doi.org/10.1016/j.njas.2015.12.003.

Darmawan, Kyuma K, Saleh A, Subagjo H, Masunaga T, Wakatsuki T (2006) Effect of green revolution technology from 1970 to 2003 on sawah soil properties in Java, Indonesia: I. Carbon and nitrogen distribution under different land management and soil types, Soil Science and Plant Nutrition 52(5):634–644. https://doi.org/10.1111/j.1747-0765.2006.00075.x.

Feledyn-Szewczyk B, Kus J, Stalenga J, Berbeć AK, Radzikowski P (2016) The Role of Biological Diversity in Agroecosystems and Organic Farming. In: Konvalina P (ed) Organic Farming - A Promising Way of Food Production. https://doi.org/10.5772/61353.

Google (2023) Google Maps [Lamongan District, East Java, Indonesia], viewed 28 February 2023. Retrieved from https://www.google.com/maps/place/Indonesia/@-2.257146,114.9289356,4.72z/data=!4m6!3m5!1s0x2c4c07d7496404b7:0xe37b4de71badf485!8m2!3d-0.789275!4d113.921327!16zL20vMDNyeW4

He H, Peng M, Lu W, Ru S, Hou Z, Li J (2023) Organic fertilizer substitution promotes soil organic carbon sequestration by regulating permanganate oxidizable carbon fractions transformation in oasis wheat fields. CATENA 221. https://doi.org/10.1016/j.catena.2022.106784.

Khumairoh U, Lantinga EA, Suprayogo D, Schulte RPO, Groot JCJ (2019) Modifying the farmer field school method to support on-farm adaptation of complex rice systems. J. Agric. Educ. Ext. 25(1):1–17. https://doi.org/10.1080/1389224X.2019.1604391.

Pfeiffer JM, Dun S, Mulawarman B, Rice KJ (2005) Biocultural diversity in traditional rice-based agroecosystems: indigenous research and conservation of mavo (Oryza sativa L.) upland rice landraces of eastern Indonesia. Environ Dev Sustain 8(4):609–625. https://doi.org/10.1007/s10668-006-9047-2.

Romasanta RR, Sander BO, Ma YKG, Alberto C, Gummert M, Quilty J, Nguyen VH, Castalone AG, Balingbing C, Sandro J, Correa T, Wassmann R (2017) How does burning of rice straw affect CH4 and N2O emissions? A comparative experiment of different on-field straw management practices. Agric Ecosyst Environ 239:143–153. https://doi.org/10.1016/j.agee.2016.12.042.

Suharjo B, Ahmady M, Ahmady MR (2016) Indonesian Consumers' Attitudes towards Organic Products. Advances in Economics and Business 4(3):132–140. https://doi.org/10.13189/aeb.2016.040303.

Uma Khumairoh Researcher and Lecturer in the Integrated Organic Farming Systems Research Centre, Faculty of Agriculture, Brawijaya University, and a guest researcher and lecturer at Wageningen University. She specializes in agroecology, complex agriculture design, bio-cultural diversity, and farmer community development.

Rochmatin Agustina Researcher and Lecturer in the Agricultural Faculty, University of Muhammadiyah Gresik and a PhD student at Brawijaya University. Her specialization is in organic plant production.

Euis Elih Nurlaelih Researcher and Lecturer in the Integrated Organic Farming Systems Research Centre, Faculty of Agriculture, University of Brawijaya. She specializes in Indonesian traditional home gardens and ecological design of agriculture.

Dewi Ratih Rizki Damaiyanti Researcher and Lecturer in the Integrated Organic Farming Systems Research Centre, Faculty of Agriculture, University of Brawijaya. Her specialization is in plant ecology and plant production.

Adi Setiawan Researcher and Lecturer in the Integrated Organic Farming Systems Research Centre, Faculty of Agriculture, University of Brawijaya. He specializes in tree cultivation and plant ecology.

Jeroen C.J. Groot Associate Professor at the Farming Systems Ecology group of Wageningen University, the Netherlands. He specializes in farming systems analysis and agroecology.

Chapter 14
Synthesis: Business and Biodiversity in the Context of Socio-Ecological Production Landscapes and Seascapes (SEPLS)

Maiko Nishi, Suneetha M. Subramanian, and Philip Varghese

Abstract This chapter synthesizes the findings from the 12 case studies presented in this volume. It is designed to answer the following questions: (1) How can we conceptualize the connections between business and biodiversity in the context of managing socio-ecological production landscapes and seascapes (SEPLS)?; (2) How can we measure, evaluate, and monitor impacts and dependency of businesses on biodiversity and nature's contributions to people through managing SEPLS?; and (3) How can we address challenges and seize opportunities in SEPLS to effectively manage business impacts on biodiversity, ecosystems, and human well-being? In addition to addressing these questions based on the synthesis of the case study findings, the chapter provides recommendations for policymakers and other stakeholders to ensure that the processes and outcomes are ecologically and socially sound and equitable in promoting more sustainable businesses. These recommendations are examples of ethical, equitable, and actionable ways of ensuring that businesses are sustainable while conserving biodiversity. In that way, businesses can contribute to achieving multiple sustainable development goals (SDGs) simultaneously,

---

Contributing authors to this synthesis chapter include: Rima Alcadi, Judy Cadingpal Baggo, Nadia Bergamini, Agnès Bernis-Fonteneau, Paolo Colangelo, Florence Mayocyco Daguitan, Emilio Rafael Diaz-Varela, William Dunbar, Fatima Ezzahra, Marco Frangella, Aristide Chacgom Fokam, Guido Gualandi, Rebecca Gurung, Hsiu Mei Huang, Shyh-Huei Hwang, Devra I. Jarvis, Uma Khumairoh, Guanqi Li, Yoji Natori, William Olupot, Anil Kumar Nadesa Panicker, Dambar Pun, Samuel Pun, Elisabetta Rossetti, Sebastian Orjuela-Salazar, Paola De Santis, Lika Sasaki, Krishna Gopal Saxena, Ye Shen, Muhabbat Turdieva, and Amanda Wheatley.

M. Nishi (✉) · S. M. Subramanian
United Nations University Institute for the Advanced Study of Sustainability (UNU-IAS), Tokyo, Japan
e-mail: nishi@unu.edu

P. Varghese
United Nations University Institute for the Advanced Study of Sustainability (UNU-IAS), Tokyo, Japan

Akita International University, Akita, Japan

M. Nishi et al. (eds.), *Business and Biodiversity*, Satoyama Initiative Thematic Review, https://doi.org/10.1007/978-981-97-7574-3_14

including not merely those specific for life below water and on land but also those related to poverty reduction, food security, good health and well-being, gender equality, quality education, capacity development, employment, climate action, responsible consumption and production, and partnerships and institutional development.

Keywords Socio-ecological production landscapes and seascapes · Landscape approaches · Business · Biodiversity · Sustainable development · Science-policy-practice partnership

1 Biodiversity, Business, and SEPLS: How Do They Connect?

The connections between biodiversity and business are primarily based on the various uses of biodiversity and the types and extent of such uses by people. On one hand, the connections relate to the dependency of businesses on ecosystems, ecosystem services, and diverse biological resources for their activities; and, on the other, they are about the impacts of businesses (whether positive or negative) on the diversity of resources and ecosystem integrity. In this volume, we have considered the nature of these interactions in the context of socio-ecological production landscapes and seascapes (SEPLS), and how we can leverage specific areas within these interactions to foster the ideal vision of living in harmony with nature. We synthesize our findings in this chapter.

1.1 SEPLS Products and Practices That Rely on Biodiversity

SEPLS are predominantly sites of primary production of various products and other ecosystem-based service activities (e.g. ecotourism), which in turn are dependent on the diverse resources and multiple functions that the varied ecosystems within the SEPLS provide.

The various business options commonly found in these contexts include:

- SEPLS products that qualify as commodities of long or short value chains: When value chains are formed, interrelationships are created across different stakeholders who may operate within and outside SEPLS (examples include supply of resources to intermediary processors and other business entities and direct sale of products to consumers). Demand for these products in value chains is fairly consistent, especially for those products that are used for staple consumption (e.g. fish, food grains, forestry products like medicinal plants, fuelwood, and timber).
- Products from SEPLS that qualify as expressions of art: These products such as crafts are of unique origin, are representative of specific cultural relationships

with ecosystems, and have niche markets. The demand for such products, even if not consistent, helps sustain the knowledge, skills, and practices associated with specific resource use (e.g. Chaps. 8 and 9) and may favour equity and social integration (e.g. Chap. 6).

- Ecotourism and such related service-oriented activities that depend on the flourishing of biodiversity and ecosystems: These businesses promote retention of traditional production practices, knowledge, diets, and landscape management and induce a strong sense of identity and place. See, for instance, chestnut forests in Spain (Chap. 3), white truffle foraging in Italy (Chap. 4), ecotourism enterprise in Nepal where traditional rice varieties are also patronized by tourists (Chap. 5), and panela production in the community sugar mill enterprise promoting ecotourism in Colombia (Chap. 11).
- Cultural activities: In the SEPLS context, the notion of culture is strong in the utilization of resources and in transactions between different actors. It is important to understand in different contexts how culture could drive business in terms of production, value addition (e.g. chestnut fruits and derived products in Chap. 3), consumption (e.g. local demand for specialized rice in China as in Chap. 12), patronage (e.g. traditional textiles in Chap. 9; traditional craft brooms in Chap. 5; and developing safe farm inputs based on traditional worldviews of doing no harm as in Chap. 9) and, how it impacts biodiversity.

1.2 *How Do Businesses Affect SEPLS?*

- In addition to business decisions made within, SEPLS are also affected by both production and land or sea use decisions made by external actors. It has often been the case that SEPLS activities are impacted by government policies (e.g. restrictions on native tree felling that adversely affected benzoin gum tree populations and further, the livelihoods of communities in Viet Nam as in Chap. 2). They are also affected by other introduced economic activities including those taking over existing activities (e.g. replacement of chestnut forests by fast-growing forestry species in Spain as in Chap. 3) or by activities and decisions arising from competition over the use of land, water, or other resources (e.g. agro-industrial practices limiting and destroying the traditional foraging areas for white truffles in Italy as in Chap. 4). These have often increased the vulnerability and uncertainties faced by SEPLS residents and actors, especially those sections of society with lesser power or influence on decisions (e.g. Cameroon where land grabbing for agro-industrial activities has rendered the SEPLS-dependent communities almost helpless and set the stage for securing their land rights as in Chap. 10).
- Businesses could also engage in the production of cheaper substitutes to SEPLS products in high demand (e.g. the truffles industry in Chap. 4) with negative impacts on traditional production in the SEPLS.

- It is also seen that awareness of the quality of SEPLS products depends on the state of biodiversity and traditional knowledge and motivates businesses to keep sustainably using raw materials produced in SEPLS (e.g. the demand for '*soutos*' chestnuts in Spain as in Chap. 3).

1.3 How to Strengthen SEPLS-Business Interconnections for Better Functionality

- *Increase consumer awareness*: As innovation is driven mainly by demand, consumer literacy that translates to demand for sustainably produced products is important (inclusive of equitable and fair-trade practices). The role of both public (e.g. administrative bodies) and non-public (e.g. environmental or cultural NGOs, local communities) organizations may be important in this respect (e.g. Chaps. 3 and 8).
- *Design and implement landscape impact assessments and monitoring protocols*: This will enhance traceability and transparency of flow of resources and chain of custody of resources across the value and supply chains. It also facilitates horizontal and vertical cooperation between planning institutions which helps to ensure a multi-scale approach to landscape management. This process for instance could build on concepts like watershed management where implementation responsibilities are clearly demarcated among actors at various scales of operation.
- *Design biocultural community monitoring protocols*: These are terms of engagement based on the biocultural diversity and contexts of the communities that develop them. This can help SEPLS actors deepen their understanding of their social and ecological assets, envision how they would like to secure their well-being, and further clarify terms on which they can engage and relate with other businesses.
- *Employ certification schemes*: The certification schemes needed are mainly those that promote businesses suitable to local culture and knowledge and biodiversity-positive products, promote sustainable use of resources like soil, land, and water, and at the same time ensure that these are affordable (e.g. BioTrade Principles and Criteria[1] as in Chap. 2).
- *Use the momentum of enabling policy and business environments*: Enabling policy and business environments could range from intergovernmental policies (such as carbon neutral development, promotion of nature-based solutions, and incentives to sustainable businesses), and business policies (such as environmental and social safeguards that incorporate biodiversity and social concerns) to

[1] Please note that while the BioTrade Principles and Criteria have been adapted to certification schemes, they are guidelines and do not involve processes such as assurance, auditing, etc.

people-led business organizations (e.g. farmers' producer organizations (FPOs) in India as in Chap. 6; see also Natori et al. 2023).

- *Understand that institutions can play an important role in realizing enabling conditions*: Examples of such institutions are farmers' producer organizations, governments, and international frameworks. The market is also an important institution, but it must be focused on creating positive outcomes for both businesses and biodiversity.
- *Advocate for keeping the prices of ecologically sound products affordable to ensure equitable access and normalizing of sustainable practices*: This does not imply unprofitable pricing mechanisms, rather ensuring the affordability of sustainably produced SEPLS products for all social groups. This also does not preclude the possibility of value-added products that are branded and sold for premium markets, which are encouraged.
- *Understand that education, awareness raising, and capacity building for communities and business managers is important for SEPLS revitalization and sustainability*: Examples include the Meister programme in Ifugao, Philippines (Chap. 7), training related to panela production and impacts in Colombia (Chap. 11), and communication campaigns by truffle hunters' associations (Chap. 4).
- *Build effective partnerships between business communities and farmers and other primary producers or entrepreneurs in SEPLS to foster healthy bio-partnerships*: In addition to fostering healthy bio-partnerships, this will help in the design of more sustainable production systems (e.g. cropping patterns that promote diversity and ecosystem functions such as soil fertility in Chap. 13; and partnership between different stakeholders involved in the production of panela as in Chap. 11).
- *Brand and personalize products for consumers with regard to SEPLS*: This implies branding products or services from SEPLS as having followed specific sustainability criteria, thereby increasing their market value (e.g. the Protected Geographical Indication 'Galician Chestnut' (Indicación Xeográfica Protexida 'Castaña de Galicia') in Chap. 3).

2 Measuring, Evaluating and Monitoring Business Dependency on Biodiversity and Nature's Contributions to People in SEPLS

Measuring and monitoring business dependency and impact on biodiversity is a challenge due to the ecological, socio-political, and cultural complexities associated with supply and value chains as well as the complexities of the businesses themselves (e.g. confidentiality agreements, trade secrets). Yet, the case studies of this volume provide rich methodological implications for monitoring and evaluating such dependency and impacts from the perspective of the supply base for business

Fig. 14.1 Keywords (represented in size) related to measuring, evaluating, and monitoring business dependency and impact on biodiversity and nature's contributions to people in SEPLS

activities. The word cloud given below (Fig. 14.1) highlights the main keywords[2] that stood out in the case studies regarding monitoring and evaluation (M&E). The aim of this section is to highlight the principles, indicators, tools, and methods that can be used in measuring, monitoring, and evaluating business dependency and business impacts on biodiversity in the context of SEPLS management.

2.1 *Principles for Monitoring and Evaluating Business Dependency and Impact on Biodiversity*

Each business interacts with biodiversity in its own way, having unique impacts (the positive or negative consequences of business activities on biodiversity) and dependencies (the ways business relies on and uses biodiversity for its activities). Despite the diverse ways businesses relate to biodiversity, there is a general agreement among the case study authors on the three broad stages of business development and their links to biodiversity, that is, planning, implementation (business introduction), and post-implementation stages. Although there are cross-cutting methods, for the most part different methods are relevant and applicable to each stage. Certain principles and criteria are pertinent to identifying appropriate methods concerning business dependency and impact on biodiversity. They are also relevant in addressing the larger issues of conservation, sustainable use, and benefit sharing of biodiversity

[2] These keywords were identified at a workshop held on 5–7 July 2023 where case study authors and reviewers convened (see preface of this volume).

through the M&E processes. The below mentioned are some of the general principles that span across all three stages.

- Clear objective of the M&E: Be it to monitor the impacts for internal use, to document changes for public disclosure, or to ensure the trust of the community, etc., it is important to start the consideration from this purpose and that the appropriate indicators are identified based on the purpose (Natural Capital Coalition 2016).
- Mix of both quantitative and qualitative methods: Combining economic valuation of biodiversity and ecosystem services with quantitative and qualitative evaluation of social impact can build an overall picture of the human interactions with ecosystems. For example, in the case study on chestnuts (Chap. 3), the qualitative approach helped to unveil the relationships between some of the trends detected by quantitative methods, such as the composite effect of producer ageing, climate effects as droughts, and new exotic pests, which have together discouraged the producers from continuing in the activity.
- Interdisciplinary and/or transdisciplinary approaches: These approaches allow for identifying indicators that are locally sensitive, flexible, and comprehensible for the people involved in SEPLS management (e.g. several complementary methodologies are utilized in Chaps. 6 and 11). It also means the synchronization of different methodologies: all indicators (science-based, cultural, and qualitative) should be synchronized across different methodologies and available frameworks. In this respect, it is relevant to establish sufficient and salient linkages between the indicators on business and those on biodiversity, as M&E activities are often faced with the challenge of dealing with unclear or only implicit linkages between business and biodiversity indicators.
- Transversal communication and awareness building: By monitoring changes, companies can continuously evaluate and act on the findings to meet their desired objectives. To realize this, support of regulators and local communities is important to identify and minimize environmental risks. In this regard, continuous educational and awareness programmes should be designed from the initial stage of M&E to identify and localize the usage of indicators to maximize the ownership of the local people and applicability of the indicators to the local people.
- Free, prior, and informed consent (FPIC): This is needed as it helps to promote bottom-up development models, and for external actors, the FPIC is a mandatory provision to consult and seek consent from the local community before initiating plans for business introduction to SEPLS. To ensure this process, workshops using the Indicators of Resilience in SEPLS are an effective way not only to guarantee the FPIC process, but also to understand the physical and social aspects of the landscape or seascape and elucidate diverse views around key issues among the stakeholders (which may help lead to consensus). The indicators enlist the key SEPLS features that contribute to resilience, whereas they can complement other existing indicator frameworks focused on community well-being, traditional knowledge, and landscape productivity. A community consultation process drawing on the resilience indicators can enable communities to

develop strategies to adapt to change through innovations and in sustainable use of biodiversity.

- Participatory approach: This approach facilitates the integration of local and traditional knowledge, local customs, and social and cultural contexts into business through sustainable practices in SEPLS (e.g. Chap. 6). In this connection, the citizen science approach helps in collecting data and monitoring the indicators. The Indicators of Resilience in SEPLS may be used in this regard as well.
- Collaborative initiatives: Businesses could partner with entities with appropriate expertise, such as NGOs, academic institutions, scientific bodies, international organizations, and/or consultants, for assistance in effective and efficient biodiversity strategy development, monitoring design, and data collection and analysis (e.g. the collaboration of academia and NGOs for the development of the collective panela production business and other initiatives such as tourism in a very fragile subxerophytic ecosystem in Chap. 11).
- Social auditing: This is an evidence-based approach that prioritizes and safeguards social and ethical principles and bridges the gap between resource utilization and effective and efficient techniques and processes that enhance the business accountability and transparency of ventures in SEPLS.

2.2 *Indicators for M&E of Business Dependency and Impact on Biodiversity*

Tangible and intangible benefits and losses resulting from business activities directly or indirectly associated with SEPLS management occur in the ecological, cultural, economic, and social dimensions. A thorough understanding of what the benefits and losses are in the specific SEPLS context (i.e. what to be measured) is an important requirement for setting the indicators and selecting tools for M&E. Adherence to the principles presented in the previous section, enhancing benefits, and minimizing biodiversity losses from businesses require the adoption of context-specific and relevant indicators. Various indicators are used depending on the SEPLS context and existing knowledge. The text below presents and discusses some of the key reference points for setting the indicators.

Biodiversity and Ecology The case studies presented in this volume as well as existing in the literature show that the higher the biodiversity, the higher the resilience and the lesser the risk of climate change vulnerability (Weiskopf et al. 2020). It is also evident from these sources that there is a significant quantitative difference in terms of biodiversity between carefully managed forests (with the higher level of biodiversity) and forests that are intensively used or have been transformed for economic performance (with the lower level of biodiversity) (Bernes et al. 2015). An increasing number of companies are addressing biodiversity issues at the core of their business. In business, focus is required to balance both the upstream and downstream of the supply chain involving materials, activities, people, and information. For example, Kering (a French-based multinational luxury fashion company) has performed a natural capital assessment to identify the magnitudes of natural capital impacts at each tier in the supply chain (Kering 2021). Based on the assessment findings, Kering also developed (and is implementing) supply chain standards (Kering 2022) to reduce its impacts and develop biodiversity strategies to regenerate farmland and conserve critical natural habitats (Kering 2023). Kering, in collaboration with Conservation International, also launched the Regenerative Fund for Nature, with the aim of bringing one million hectares of croplands and rangelands under regenerative agriculture by 2025.[3]

Knowledge and Culture While adapting to environmental change, such as climate change, SEPLS management maintains the connections to traditional and cultural roots through citizen-centric approaches. The social and knowledge integration across diverse values and perspectives helps to identify and understand human-nature interactions and to better understand the trade-offs. Modern science often contributes to explaining the mechanisms of traditional ecological knowledge and customary practices. For example, the combination of traditional knowledge and modern science and technology has helped the 'renewal' of customary practices to overcome the challenges imposed by the spread of fungal diseases in the chestnut orchard recovery projects (Díaz-Varela et al. 2018).

Availability of Livelihood Alternatives A range of viable alternatives and new opportunities of livelihood for community members allow for their active engagement in sustainable business practices in SEPLS. Identifying such alternatives and opportunities can be done through local people's participation and their goal setting could be built on a rich and diverse portfolio of biodiversity. Local adoption of sustainable business practices would entail sustainable customary use of biodiversity and employment benefits for the local community. Such practices would facilitate the engagement of local stakeholders.

Social Integration Social incentives and competence to effectively manage SEPLS improve with greater participation of local people. Businesses developed within SEPLS can benefit from the establishment of formal or informal networks of producers which rely on similar inputs (e.g. shared knowledge, opportunities, and market trends) from the network. For example, the case from India (Chap. 6) shows how cultural values favoured equity, environmental sustainability, and social integration (i.e. various communities sharing societal beliefs, norms, values, and customs as a whole).

Enhancing Crop Yield Crop productivity, quality of yield, and the amount of use of chemicals are clear and strong indicators of management of SEPLS (e.g. see the

[3] https://www.kering.com/en/sustainability/safeguarding-the-planet/regenerative-fund-for-nature/

case of complex rice systems in the Lamongan district of Indonesia as discussed in Chap. 13).

Objective and Subjective Measurements While it is indispensable to conduct assessments across various scales based on objective quantitative measurements, it is also important to have indicators based on subjective evaluations (e.g. Chap. 13). The biophysical information (e.g. species richness) should be supplemented with environmental, social, economic, cultural, and governance dimensions (such as human health, community-based landscape governance, women's knowledge, and traditional knowledge—often evaluated by qualitative and subjective measurements), and vice versa, to harness a holistic picture in M&E.

Human Well-being People's happiness and well-being indicate dimensions that are dependent on and associated with the multifunctionality of SEPLS and can be measured by multiple indices (such as access to water, food security, health, and livelihoods) for M&E. The case study on the Pidlisan Tribe of Sagada in the Mountain Province of the Philippines (Chap. 8) shows how the social enterprise not only produced positive results in terms of soil fertility, enhanced yield, and biodiversity, but also helped workers in terms of time management and their well-being. Workers also benefited from the platform they created for building and exchanging knowledge and enhancement of skills.

Flexible Indicators Based on Interdisciplinary and Transdisciplinary Knowledge Different circumstances and geographies imply the need for the flexible and adaptive capacity of indicators. Such flexible indicators build on the interdisciplinary and transdisciplinary nature of knowledge, while allowing for the use of existing available and accessible knowledge to make the indicators useful. To make various indicators easily understood and continuously used at the local level, capacity development initiatives (e.g. training and facilitation of local assessment work) would be helpful. The process of adjusting indicators to the local context and integrating local innovations and sustainable use of biodiversity should carefully attend to the protection of vulnerable groups from the risks associated with business. In this way, the M&E process would lead to the development of simple, clear, and intuitive, but at the same time multidimensional indicators to be effectively used in accordance with the local context.

2.3 Tools and Methods for M&E of Business Dependency and Impact on Biodiversity

In many of the case studies, a combination of different methods is used based on time-tested practices specific to each SEPLS, where indicators are chosen in accordance with the assessment type, local contexts, and M&E purposes and objectives.

Furthermore, the studies suggest that repeated monitoring (i.e. assessments are conducted throughout, including pre- and post-introduction of business to an area) would allow for studying the impact and dependency of a particular business on biodiversity in the area. Despite the varied choice of methods and tools depending on the circumstances, the following are several tools and methods which could not only be useful for M&E but also promote sustainable business practices in the context of SEPLS.

Integrated and Comprehensive Assessment Frameworks The ecosystem services framework could be used to identify business dependencies and impacts on biodiversity with a focus on how the ecosystem functions for human well-being, by organizing data and information concerning different types of ecosystem services (i.e. provisioning, regulating, cultural, and supporting services) and habitats. Similarly, the livelihoods framework addresses different aspects of business dependencies and impacts on biodiversity, including human, ecological, social, physical, and financial assets to enhance livelihood opportunities or minimize the risks. Furthermore, the Natural Capital Protocol (Natural Capital Coalition 2016) is a useful framework enabling organizations, including those in business and finance, to identify, measure, and value their direct and indirect impacts and dependencies on natural capital for their decision making. This model is also followed by the LEAP (Locate, Evaluate, Assess and Prepare) approach, an integrated approach developed by the Task Force on Nature-related Financial Disclosures (TNFD) to enable various types of organizations to integrate and assess nature-related issues.

Biodiversity Assessment Tools These are the tools that can be used to assess the biodiversity that businesses depend on and their impacts and can be applied to various levels of business activities ranging from product development and site maintenance to supply chain analysis. Examples include the Diversity Assessment Tool for Agrobiodiversity and Resilience (DATAR) (see Box 14.1), Biodiversity Impact Metric, Biodiversity Performance Tool and Monitoring System (BPT), Species Threat Abatement and Recovery (STAR), Integrated Biodiversity Assessment Tool (IBAT), Lasting Initiative for Earth (LIFE), and Science-Based Targets for Nature (Katic et al. 2023). Despite the general usability of these tools, a careful choice of appropriate tools and methods for biodiversity assessments could be made in accordance with M&E purposes as individual businesses rely on biodiversity in their specific contexts and interests, which then have certain (positive and/or negative) consequences for biodiversity. Business dependency on biodiversity can be rather indirect in many cases as it is more clearly captured through ecosystem services that are supported by biodiversity (e.g. the volume of water used in business), while such dependency (e.g. water use) makes certain impacts on biodiversity (e.g. water loss, habitat change) which can be measured by some of the biodiversity assessment tools. Nevertheless, several businesses and commercial activities like ecotourism or nature-positive tourism, as in the case of the Panchase Region of Nepal (Chap. 5), directly depend on high levels of biodiversity (Brandt and Buckley 2018).

Box 14.1 Diversity Assessment Tool for Agrobiodiversity and Resilience (DATAR)[4]

Farmed intraspecific diversity is still widely found in smallholder production systems that are SEPLS. The Diversity Assessment Tool for Agrobiodiversity and Resilience (DATAR; Free Open Source www.datar-par.org) is a decision-making tool supporting the improvement of local communities' livelihoods and benefits from the use of their local intraspecific diversity and the restoration of ecosystem health. The DATAR methodology offers an opportunity to assess the presence, distribution, access, and use of agrobiodiversity to feed and restore our planet through SEPLS management, in the form of crop varieties, livestock breeds, and aquatic farmed types. It has been developed using numerous case studies around the world and more than 25 years of experience in the field. Agrobiodiversity in production systems is the result of farmers' actions and choices. DATAR allows for a better understanding of humanly preserved and managed agrobiodiversity and its distribution as well as associated knowledge and practices. Information is collected through surveys on these issues linked to the numerous factors impacting agricultural production systems including agroecological zones, socio-economic characteristics, market aspects, and policies and regulations of the area studied. The system connects results from surveys and empirical data at the farm and landscape levels to a heuristic decision-making framework with community goals and constraints faced by farmers. It then provides adapted portfolios of diversity-related interventions to be implemented. Interventions range from characterization of materials and identification of essential traits to in situ improvement of local materials and adapted or new business opportunities for agrobiodiversity conservation (e.g. the use of geographic indications or quality-assured producer and product, on-farm agrobiodiversity ecotourism, market creation and promotion for landraces and for indigenous, and locally adapted breeds and their products). Monitoring and evaluation with DATAR allows for measuring the achievement of diversity-related objectives set by project coordinators and farming communities, and achievement of global environmental benefits on environmental, economic, social, and health aspects.

[4]The development of the DATAR WEB Portal and the DATAR APP was made possible through the generous support of the Global Environmental Facility (GEF)—International Fund for Agricultural Development (IFAD): Cross-cutting capacity building, knowledge services, and coordination project for the Food Security Integrated Approach Pilot Program—GEF project 9140. For more details, see also Bernis-Fonteneau et al. (2023).

Source: www.datar-par.org

Landscape or Seascape Mapping An assessment of land and sea uses, taking into consideration all physical, human, and natural elements, forms the basis of this mapping. Although a variety of methodologies have been developed and adopted for landscape or seascape classification and identification, the central idea remains to acquire a holistic picture of land or sea use, socio-cultural aspects, land/seascape characteristics, physical elements and properties through visual perception, geo-ecological studies, and statistical analysis (Simensen et al. 2018).

Certification Schemes Developing a certification system for SEPLS management in accordance with the local context helps with making business activities more sustainable. The Participatory Guarantee System (PGS) of the Ministry of Agriculture and Farmers Welfare, Government of India, is an example of locally-focused quality assurance. This certification was developed based on active participation of stakeholders, building on trust, social networks, and knowledge exchange among the stakeholders. Other examples are Rainforest Alliance certification, Roundtable on Sustainable Palm Oil (RSPO) certification, Forest Stewardship Council (FSC) certification, and the Union for Ethical BioTrade (UEBT) standard. The Climate, Community & Biodiversity Standards can also be used with carbon accounting schemes to ensure the delivery of biodiversity and community benefits from land-based climate change mitigation projects (Climate, Community & Biodiversity Alliance 2017). The cost of getting certified is a genuine concern for small businesses. Simple, affordable alternatives could be developed to encourage biodiversity-sound business development for small and medium enterprises.

Stakeholder Identification and Analysis Across all the phases of planning, implementation, and post-implementation, the involvement of stakeholders is crucial. At the beginning of the business development process, businesses must identify relevant stakeholders and analyse their interests, concerns, and roles, as well as the potential impacts the business activities can have on them. This is an important part of the social safeguard. Multi-stakeholder platforms would encourage collective learning and consensus building and create shared spaces for negotiation and conflict resolution, for example, integration of local exporters and producers into global value chains through BioTrade (Chap. 2) and stakeholder identification, characterization, and prioritization for analysis (Chap. 11).

Visual Assessment This is a primary tool which is often used by farmers to account for the species richness, evenness, and relative dominance. Furthermore, in addition to using remote sensing and drones, various tech-driven solutions have emerged such as a new user-friendly programme VBioindex, which was developed for biologists and ecologists to draw on various indices for better visual representations of the aforementioned features of species in the form of simple charts (Yu and Yoo 2015).

Multiple Surveys Baseline, midline, and endline surveys are adopted to evaluate the impact as well as effectiveness of the interventions made. This also helps in accountability, milestone setting, and overall learning or knowledge building through M&E.

Results-Based Payment Systems These systems are being tested in mountain regions of Spain and other parts of Europe to preserve ecosystems that people depend upon for traditional livestock husbandry and agricultural practices and to arrest habitat degradation (Pinto-Correia et al. 2022; Moran 2020). Specific examples were put in practice by the recently finished 5-year LIFE project ‘LIFE in Common Land’, known as the Layman’s Report (Blanco-Arias et al. 2023). Similar examples of payment for ecosystem services programmes include the REDD+ and Ecuador’s Socio Bosque Program.

Rewards System A participatory rewards system by which incentives are given to people involved in the M&E process is being promoted. The economic incentives that can derive from this system could range from management or co-management of property, establishment or improvement of biodiversity markets, tax relief, subsidies, and conditional bonds for restoration. Other livelihood-supporting indirect incentives include rural development and income diversification. These incentives are usually quick and easy to implement; however, rewards systems may lack scientific credibility unlike other above-mentioned rigorous scientific tools.

3 Ways Forward: Leveraging Landscape Approaches for Sustainable Businesses

Landscape approaches are long-run processes where a wide range of stakeholders collaborate to balance multiple objectives, priorities, and needs on a landscape or seascape scale. With multi-stakeholder collaboration involving local communities, businesses and industries, and other actors associated with supply chains, they help to minimize trade-offs and maximize synergies to the greatest extent possible in promoting biodiversity-friendly businesses that ensure sustainability. Among others, businesses play a critical role in creating and facilitating a supply chain connecting suppliers (producers) to end users (consumers), through which trade-offs and synergies occur between various associated elements. In the context of SEPLS management, such trade-offs and synergies can be identified largely at the two levels: (1) within a landscape or seascape scale and (2) beyond the focal landscape or seascape scale (along a supply chain and beyond).

3.1 Trade-Offs and Synergies Within a Landscape or Seascape

Trade-offs arising from competing needs and demands and thereby contradictory practices often occur in intimate connection with various natural and socio-economic resources within a land/seascape. For instance, a successful increase in crop yields or harvest with modern (and often expensive) technologies may undermine food security in the long run or the quality of products that had been ensured through traditional practices (e.g. Chaps. 4, 6, and 9), while depressing traditional management practices and associated knowledge and sometimes jeopardizing the local livelihoods based on them (e.g. Chaps. 3 and 10). Beyond the cases compiled in this volume, literature also shows similar examples that are manifested within a landscape or seascape. Examples include an increase in multiple competing demands for shea nut trees (e.g. shea butter production, charcoal making) pressuring the natural regeneration of the trees (Buyinza and Okullo 2015), and a shift from traditional nomadic pastoralism (which alleviates pressure on natural resources) to sedentary land use (which requires massive use of supplemental feed and leads to overgrazing (Chap. 6)—see also (Martínez-Valderrama et al. 2018).

However, synergies positive to multiple dimensions can also be observed within land/seascapes. The case from Indonesia (Chap. 13) epitomizes the harmonized cohabitation of different species beneficial for production as well as the sound combination of multiple production systems. In this case, additional crops are grown on bunds in the rice fields, and herded ducks are fed with leftover rice grain in the fields after rice harvesting. Also, diverse resources within a landscape or seascape (e.g. traditional knowledge, natural resources, local stewardship) are used or revived with innovative ideas, allowing for intergenerational knowledge transmission, identification of new business opportunities, conservation and sustainable use of biodiversity, and more resilient local communities (e.g. Chaps. 7, 8, 9, and 11).

3.2 Trade-Offs and Synergies Beyond the Landscape or Seascape

The above-mentioned local or regional trade-offs as well as synergies are often either derived from or amplified by external factors that are directly or indirectly associated with a supply chain. A supply chain entails multiple actors across different scales (e.g. primary producers, harvesters, manufacturers, processors, intermediaries, wholesalers, distributors, product designers, retailers, and consumers) whose activities are also supported or influenced by other stakeholders (e.g. governments, civil society, and academia) as well as broader social, economic, and political systems. These diverse stakeholders along the supply chain and beyond often have different priorities, interests, and value perspectives, which are not always commensurable but could be conflicting, giving rise to trade-offs. Government policy for industrialization or economic development may contradict local needs for livelihoods, whereas monopolization of large industries severely outpaces environmentally friendly small enterprises (e.g. Chaps. 6 and 10). Numerous cases beyond this volume show that conservation benefits can be easily compromised by the consumer's quest for affordability and comfort.

Nevertheless, synergies can also happen through beneficial interactions among these actors who have different advantages or virtues to be leveraged. For instance, businesses can take advantage of local histories and/or traditional ecological knowledge to add value to their products that are supplied through biodiversity-friendly practices (e.g. Chaps. 5, 7, 9, and 12). Such initiatives can be supported by governments through subsidies or incentive measures, whereas equitable financing can support imperative income generation particularly of those vulnerable to market fluctuations. Furthermore, social entrepreneurship can be fostered to bring about social innovation under the shared values among stakeholders, while different activities associated with a supply chain can be streamlined with useful technologies (e.g. information and communication technology) to optimize economic profit and social and environmental impacts. For instance, Kering, as a company that owns several luxury fashion brands, not only demands its suppliers (and through them, their sub-suppliers) meet its environmental standards, but also supports them to develop the necessary capacity based on the understanding that its business is only as good as what is supplied and that biodiversity is a source of inspiration to its designers.

3.3 Barriers and Enablers

There are a series of barriers and enablers of dealing with trade-offs and synergies to attain sound social and ecological outcomes through business activities, including those related to awareness and recognition, and institutional arrangements. First, awareness and recognition of business impact and dependency on biodiversity is a bottleneck in addressing trade-offs and creating synergies within and beyond a landscape or seascape. Full awareness and precise understanding of such impact and dependency allow us to take appropriate action and change business-as-usual practices for more sustainable businesses. As discussed in the previous section, various methodological challenges—including those to measure not only ecological impacts but also consequences for multiple dimensions of human well-being—hinder our complete understanding of business implications for biodiversity. This impedes appropriate decisions and actions to ensure sustainability. It is often the case that traditional knowledge has been embodied over generations through augmenting experiences and wisdom concerning sustainable use of biodiversity (e.g. Chaps. 3, 4, 8, 9, and 13). Such knowledge often underpins sustainable management practices and even helps to create new business opportunities in combination with other types of knowledge (e.g. modern sciences) for biodiversity-friendly innovation (e.g. Chaps. 5, 7, and 9). Yet, a loss or weakening of traditional knowledge and skills and their deficient transmission to younger generations have been observed across different regions around the world (Aswani et al. 2018). Conversely, the efforts in supporting intra- and intergenerational sharing of such knowledge and skills allow for promotion of sustainable businesses.

Second, institutional arrangements as well as governing processes either hamper or facilitate effective communications and collaboration to promote sustainable business practices and engage the actors associated with the supply chain directly or indirectly in such practices. Uni-directional relationships between downstream and upstream supply chain actors and communication gaps between different communities (e.g. conservationist communities vs. business communities) can inhibit mutual understanding, leading to missed opportunities for meaningful joint efforts to better handle negative trade-offs and establish positive synergies. Landscape approaches provide a framework to enable multi-stakeholder negotiation and collaboration, but governmental planning cycles may not match the framework rendered by the landscape approaches. For instance, the official planning procedures seek short-term impacts, whereas social-ecological outcomes through landscape approaches may come to the fore after long-term complex and non-linear processes. As such, even though global goals and targets have been established for climate and biodiversity towards 2030 and 2050, ecological responses to policy interventions along with climate and other environmental changes can be slow or stochastic, and it may take more decades or even centuries for ecological processes to be fully restored or conserved (e.g. late-successional habitats) (Pearce-Higgins et al. 2022).

Financial institutions and markets also serve a crucial role in informing producers and consumers among others to facilitate sustainable businesses. For example, the case study concerning truffle production in Italy (Chap. 4) suggests that if business opportunities are abundant or at least remunerative, traditional knowledge and skills for truffle harvesting could have been passed down to younger generations, contributing to sustainable practices of ecosystem management and truffle production. However, given the current dominant market mechanisms, biodiversity-friendly business practices where traditional knowledge and skills are appropriately integrated are hardly mainstream in financial and market domains.

The improvements in both institutional frameworks and awareness and recognition of all the stakeholders can go hand in hand to promote sustainable businesses. For instance, partnering or networking among local associations of producers, NGOs, and administrative bodies can help to share similar concerns and develop collective action in addressing the issues including those related to businesses. The shared visions or common objectives may allow for bringing together a broader range of stakeholders and establishing collaboration among them for biodiversity-friendly business practices. In this regard, the case study of the BioTrade initiative in Viet Nam (Chap. 2) shows the shift in the mindset among a broad range of stakeholders associated with a value chain. Its multi-stakeholder approach allowed for changing their ideas of and attitudes toward the benzoin trees from cutting down the trees for logging to conserving both the health and population of the trees so that benzoin gum can be continuously collected. This leads to sustainable business practices that entail conserving local biodiversity, supporting income generation among local communities, and fostering a sustainable value chain while enabling buyers to meet their clients' demands.

A lack of financial resources could often be an impediment to promote biodiversity-friendly business activities. Yet, there is a growing market of investment in which investors seek positive impacts on the environment and society (i.e. impact investments). For instance, the over 150 financial institutions that are signatory to the Finance for Biodiversity Pledge collectively have over EUR 21 trillion in assets under management.[5] The challenge in accessing these funds is to create projects and initiatives that can be considered for investment (as opposed to grants) that are able to demonstrate and generate returns. To tackle these challenges, a multi-stakeholder approach helps to mobilize financial and technical resources from different stakeholders and enhance their capacities to identify workable solutions and facilitate concerted efforts in promoting sustainable practices (e.g. Chap. 11).

3.4 *Steps to be Taken to Leverage Landscape Approaches for Sustainable Businesses*

Landscape approaches are place-based approaches that facilitate multi-stakeholder dialogues and enable collaborative efforts leading to sustainable solutions for both businesses and biodiversity. Grounded in local practices and experiences, the approaches help to reconcile different views and values on human-nature interactions in a given area and then enhance and extend the collaboration and partnership along and beyond the supply or value chain to mainstreaming biodiversity in businesses. This is an iterative process that involves adaptation and course correction. As not mutually exclusive but repetitive and overlapping processes, the following steps are recommended to be taken collectively by policymakers and other

[5] https://www.financeforbiodiversity.org/

stakeholders, particularly including those in the business and finance sector, to facilitate sustainable businesses:

1. Build on local stewardship to establish a common vision among stakeholders

   - Ensure recognition and raise awareness of biodiversity values among local stakeholders and mobilize them to act on their appreciation of such values for their productive activities. This can be pursued by taking advantage of local collectives (e.g. farmers associations and cooperatives).
   - Enhance education and promote communication among different stakeholders within and beyond a land/seascape scale through demonstration and peer-learning of success stories and good practices so that they can be replicated and upscaled (e.g. incorporating environment and biodiversity components in business education programmes).
   - Facilitate finding a common vision around which all stakeholders can reconcile different views and values and address all pillars of biodiversity conservation. This can be achieved by creating a common platform or communication channels to bring together different stakeholders who are directly and indirectly associated with a supply or value chain for sharing their concerns, views, and value perspectives on biodiversity and sustainability.

2. Institutionalize cross-level collaboration and review mechanisms

   - Institutionalize a participatory approach to planning and implementing biodiversity-friendly business practices at the land/seascape level while ensuring collaboration across different levels (from local, subnational, national to international) where rights of local actors are fully acknowledged and protected. In this process, new practices are innovated and tested at the local level through empowering local communities, whereas relevant institutions are built and enhanced across different levels.
   - Establish mechanisms to review and monitor supply and value chains and continuously adapt them to changes and if appropriate, upscale them to different communities and countries. In this process capacity development is crucial, for instance, with regards to impact studies, risk assessments, promotion of the concepts of adaptiveness and flexibility through training, and participatory approaches to monitoring and evaluation. This allows for ensuring accountability and transparency of businesses to collaboratively manage a risk of negative biodiversity impacts among stakeholders while taking advantage of biodiversity for businesses.
   - Build a sustainable model of biodiversity-based partnership through which the following elements can be integrated and interconnected at the local level and then scaled up: (1) recognition of biodiversity as an essential source for local livelihoods and well-being as well as businesses; (2) biodiversity-friendly business practices; (3) promotion of biodiversity-friendly trade (e.g. BioTrade); and (4) multiple benefits arising from the entire partnership. This may require long-term planning where some tools or systems such as GIS can help with identifying and evaluating biodiversity and other local resources as

well as place-based contexts, and then with sharing the information and knowledge among stakeholders for a broader partnership.

3. Mainstream biodiversity in businesses to achieve global goals for biodiversity and sustainability

   - Foster mutual trust between business and biodiversity communities through careful consideration of access and benefit sharing by hinging on the Nagoya Protocol on Access to Genetic Resources and the Fair and Equitable Sharing of Benefits Arising from their Utilization under the Convention on Biological Diversity (CBD) (e.g. business use of medicinal plants and other bioresources).
   - Ensure and support the SEPLS management by businesses to be aligned with global policy agendas, including the sustainable development goals (SDGs), Paris Agreement, Kunming-Montreal Global Biodiversity Framework, and UN Decade on Ecosystem Restoration.

4 Conclusion

SEPLS are sites of primary production activities (whether agriculture, fisheries, or crafts industries) that are dependent on biodiversity and well-functioning ecosystems. The products from these sites enter trade and businesses through markets and other business channels. These markets could be localized or be part of larger national and international commodity chains. Market demand, policy imperatives, production practices of other businesses in the land/seascape, demographic changes, and cultural priorities are some factors that influence how diverse, equitable, and aligned with sustainability are the activities within the land/seascape.

The case studies in this volume demonstrate the various ways businesses are dependent on biodiversity and the various ways different actors are dependent on each other and on the diversity of nature and integrity of ecosystems for their well-being. Consequently, activities in one sector by some actors will have (positive or negative) repercussions on others. This implies that more deliberative planning processes are required, following principles of co-learning and co-design that will help reduce trade-offs between various activities.

The cases also illustrate proactive actions and furthermore monitoring and assessment of social and ecological parameters in the SEPLS that enable building better partnerships for trade and business. This includes gaining trust and obtaining the free, prior, and informed consent of communities through transdisciplinary approaches where all potential stakeholders are engaged. They also show that when consciously designed, value addition activities (whether ecotourism, food processing, or craft products) can promote and strengthen local cultural practices and knowledge on ecosystem management and use (as in revitalizing traditional foods, crafts, diets, and other cultural resources).

Upscaling and sustaining trade and business activities in a manner that is aligned with sustainability principles hinges on building strong partnerships among different types of stakeholders involved in the value or supply chain and strengthening institutions that can facilitate and oversee business development. Decision support tools such as DATAR enable organization of several parameters related to these aspects. Certification schemes or standards and guidelines that adhere to sustainability principles, such as the BioTrade Principles, also help steer discussions and decisions that are mindful of multiple interests of different stakeholders, especially communities in SEPLS. Facilitating a proper assessment of business impacts and dependencies on biodiversity as well as raising awareness and communicating through education and training modules, peer learning, and regular stakeholder consultations should minimize the trade-offs associated with business and biodiversity. They would also enhance the likelihood of leveraging synergies that may exist such as cross-resource utilization (e.g. financial, technical, and human resources) across sectoral interests.

Given the dynamic nature of such systems, constant monitoring and review to account for contextual changes are necessary. Measures such as social auditing, visual assessments, spatial mapping, and multi-stakeholder consultations help to monitor the status of the social-ecological system.

References

Aswani S, Lemahieu A, Sauer WH (2018) Global trends of local ecological knowledge and future implications. PloS one 13(4):e0195440.

Bernes C, Jonsson BG, Junninen K, Lõhmus A, Macdonald E, Müller J, Sandström J (2015) What is the impact of active management on biodiversity in boreal and temperate forests set aside for conservation or restoration? A systematic map. Environ Evid 4, 25. https://doi.org/10.1186/s13750-015-0050-7

Bernis-Fonteneau A, Alcadi R, Frangella M, Jarvis DI (2023) Scaling Up Pro-Poor Agrobiodiversity Interventions as a Development Option. Sustainability (Switzerland) 15(13). https://doi.org/10.3390/su151310526

Blanco-Arias CA, Laborda-Bartolomé A, Cardoso-Canastra F, Lagos Abarzuza L, Díaz-Varela ER, Fagúndez-Díaz J, Díaz-Varela RA (2023) LIFE in Common Land: Management of communal land, a sustainable model for conservation and rural development in Special Areas of Conservation (Layman's Report), viewed 22 September 2023. Retrieved from https://lifeincommonland.eu/uploads/files/YDRAY-INTERACTIVO_CICA_LIFE-IN-COMMON-LAND_INFORME_ENG_compressed.pdf

Brandt JS, Buckley RC (2018) A global systematic review of empirical evidence of ecotourism impacts on forests in biodiversity hotspots. Curr Opin Environ Sustain 32:112–118. ISSN 1877-3435

Buyinza J, Okullo JB (2015) Threats to conservation of Vitellaria paradoxa subsp. nilotica (Shea butter) Tree in Nakasongola district, Central Uganda. International Research Journal of Environment Sciences 4(1):28–32

Climate, Community & Biodiversity Alliance (2017) Climate, Community & Biodiversity Standards ver. 3.1. https://verra.org/wp-content/uploads/CCB-Standards-v3.1_ENG.pdf

Díaz-Varela S, Pascual U, Stenseke M, Martín-López B, Watson RT, Molnár Z, Hill R, Chan KMA, Baste IA, Brauman KA, Polasky S, Church A, Lonsdale M, Larigauderie A, Leadley PW, Van Oudenhoven APE, Van Der Plaat F, Schröter M, Lavorel S, Aumeeruddy-Thomas Y, Bukvareva E, Davies K, Demissew S, Erpul G, Failler P, Guerra CA, Hewitt CL, Keune H, Lindley S, Shirayama Y (2018) Assessing nature's contributions to people: Recognizing culture, and diverse sources of knowledge, can improve assessments. Science 359(6373), viewed 22 September 2023. Retrieved from https://www.science.org/doi/10.1126/science.aap8826

Katic PG, Cerretelli S, Haggara J, Santika T, Walsh C (2023) Mainstreaming biodiversity in business decisions: Taking stock of tools and gaps, Biological Conservation 277:109831. ISSN 0006-3207

Kering (2021) Environmental Profit and Loss (EP&L): 2021 Group Results. Kering, Paris. https://www.kering.com/api/download-file/?path=Kering_Environmental_Profit_and_Loss_Report_2021_EN_only_09b0ab0899.pdf

Kering (2022) Kering Standards: Standards & guidance for sustainable production. Kering, Paris. https://www.kering.com/api/download-file/?path=Kering_Standards_V5_EN_68e672bc82.pdf

Kering (2023) Biodiversity Strategy: Bending the Curve on Biodiversity Loss (version 2.0). Kering, Paris. https://www.kering.com/en/sustainability/safeguarding-the-planet/biodiversity-strategy/

Martínez-Valderrama J, Ibáñez J, Del Barrio G, Alcalá FJ, Sanjuán ME, Ruiz A, Hirche A, Puigdefábregas J (2018) Doomed to collapse: Why Algerian steppe rangelands are overgrazed and some lessons to help land-use transitions. Science of the Total Environment 613:1489–1497

Moran J (2020) Policy Environment: Ecosystem services and the role of Results-based Payment Schemes (RBPS) in integrated approach to agricultural land use. In: Farming for Nature: The Role of Results-Based Payments, Teagasc and National Parks and Wildlife Service (NPWS), Dublin, Ireland, pp 250–273

Natori Y, Kharrazi A, Portela R, Gough M (2023) Nature-Based Solutions in the Private Sector: Policy Opportunities for Sustainability in a Post-Pandemic World. In: Filho WL, Ng TF, Iyer-Raniga U, Ng A, Sharifi A (eds) SDGs in the Asia and Pacific Region. Implementing the UN Sustainable Development Goals—Regional Perspectives. Springer, pp 1–23. https://doi.org/10.1007/978-3-030-91262-8_113-1

Natural Capital Coalition (2016) Natural Capital Protocol. Natural Capital Coalition. www.naturalcapitalcoalition.org/protocol

Pearce-Higgins JW, Antão LH, Bates RE, Bowgen KM, Bradshaw CD, Duffield SJ, Ffoulkes C, Franco AMA, Geschke J, Gregory RD, Harley MJ, Hodgson JA, Jenkins RLM, Kapos V, Maltby KM, Watts O, Willis SG, Morecroft MD (2022) A framework for climate change adaptation indicators for the natural environment. Ecological Indicators 136:108690

Pinto-Correia T, Ferraz de Oliveira MI, Guimarães MH, Baptista ES, Pinto-Cruz C, Godinho C, Santos RV (2022). Result-based payments as a tool to preserve the High Nature Value of complex silvo-pastoral systems: progress toward farm-based indicators. Ecology and Society 27(1):39

Simensen T, Halvorsen R, Erikstad L (2018) Methods for landscape characterisation and mapping: A systematic review. Land Use Policy 75:557–569. ISSN 0264-8377

Weiskopf SR, Rubenstein MA, Crozier LG, Gaichas S, Griffis R, Halofsky JE, Hyde KJW, Morelli TL, Morisette JT, Muñoz RC, Pershing AJ, Peterson DL, Poudel R, Staudinger MD, Sutton-Grier AE, Thompson L, Vose J, Weltzin JF, Whyte KP (2020) Climate change effects on biodiversity, ecosystems, ecosystem services, and natural resource management in the United States. Science of the Total Environment 733:137782. ISSN 0048-9697

Yu DS, Yoo SH (2015) VBioindex: A Visual Tool to Estimate Biodiversity. Genomics & Informatics 13(3): 90–92

Maiko Nishi Research Fellow at United Nations University Institute for the Advanced Study of Sustainability. Her research interests include social-ecological system governance, local and regional planning, and agricultural land policy.

Suneetha M. Subramanian Research Fellow at United Nations University Institute for the Advanced Study of Sustainability. Her research interests include biodiversity and human well-being with a focus on equity, traditional knowledge, community well-being, and social-ecological resilience.

Philip Varghese JSPS-UNU Postdoctoral Fellow at Akita International University, Japan and United Nations University Institute for the Advanced Study of Sustainability. His research interests include tourism and development politics, indigenous communities, and sustainability.